Turning Barriers Into Bridges

Turning Barriers Into Bridges:
The Inclusive Use of Information and Communication Technology for Churches in America, Britain, and Canada

by John Jay Frank, PhD
July 26, 2015
Updated September 12, 2021

Published by Minstrel Missions LLC
minstrelmissions@gmail.com
www.minstrelmissions.com

"BUT NOW BRING ME A MINSTREL . . .

But Now Bring Me A Minstrel . . .
(2 Kings 3:15, KJV) or
(Ezekiel 33:32 KJV)

* * * * * * * *

Turning Barriers Into Bridges:
The Inclusive Use of Information and
Communication Technology for Churches in
America, Britain, and Canada

ISBN: 978-1-887835-21-3 Large Print Edition
Published by Minstrel Missions LLC
minstrelmissions@gmail.com
www.minstrelmissions.com

Available in paperback in 12 point Arial font and large print, 18 point font and in a Kindle format.

Keywords and book categories:
1. Ministry and Evangelization, 2. Technology and Church Growth, 3. Missiology, missions, and Information and Communication Technology (ICT), 4. Disability and the Bible, 5. Inclusion or disability discrimination in church, 6. Bible study, 7. Christian nonfiction.

* * * * * * * *

Preface:

My thanks go out to all those who read and commented on this book. Some, especially those with one of the hidden impairments I write about, a vision or hearing impairment, or limited reading ability, were ecstatic at this expression of their personal experiences, concerns, and the real and reasonable solutions available to churches. Some suggested additional concerns they wanted me to include, but those, while also important, were not specifically related to the book's focus, that is, the use of information and communication technology.

Some readers who had not experienced the barriers we often create in church did not know how difficult it can be to discuss and initiate even easy-to-do changes. They were surprised by the many opportunities now available to us to include those we may have excluded. Please consider that as a result of this book, one volunteer was asked why she always made the print on their weekly church bulletin so small, since some people could not see well enough to read it. She replied, "I never even thought of that." Perhaps now there is a hope and beginning of change.

This book is dedicated to those who presently are able, and those who are being disabled and especially those who work to prepare a gathering place for all of God's people. – John Jay Frank

You are the light of the world.
A city set on a hill cannot be hidden.
(Matthew 5:14)

My Gracious Master and my God,
Assist me to proclaim,
To spread through all the earth abroad
The honors of Thy name.

Jesus the name that charms our fears;
That bids our sorrows cease,
'Tis music in the sinners ear,
'Tis life, and health, and peace.

He speaks and listening to His voice
New life the dead receive,
The mournful broken hearts rejoice,
The humble poor believe.

He breaks the power of canceled sin,
He sets the prisoners free;
His blood avails for all our sin,
His blood availed for me.

Hear Him ye deaf, His praise ye dumb,
Your loosened tongues employ;
Ye blind behold your Savior come,
And leap ye lame for joy.

O for a thousand tongues,
Charles Wesley, 1738, Public domain.

Contents

Note from the author: The look of a book may depend on where it is printed. The colors on this cover attempt to demonstrate the contrast mentioned in chapter 10, but colors for ink printing vary and cannot accurately represent computer projection. The text is 18 point Arial font, with 20 point chapter titles, and 16 point page numbers headers and references. It is also available as a 12 point Arial font paperback, and in a Kindle format, which is a somewhat accessible eBook with audio capability. This book presents many standards, but does not give instructions for using programs.

1. Introduction

For better and for worse, technology alters the way we share our messages in church. I can remember after a youth group meeting taking the typewritten bulletin to a local print shop to be professionally printed. Back then, it was difficult or impossible for some people to read the small print in a Bible, a hymnal, or other materials in church. Years later, churches bought photocopy machines, overhead projectors, computers, and printers. Then along came computer projectors, LCDs, the Internet, cell phones, digital cameras, and homemade DVDs. Today, these are routinely used in church and outreach ministries by people with some knowledge of how to use computers.

I recently shared with an optometrist some of the many ways information and communication technology (ICT) is used in churches which make participation impossible for many people with low vision. She replied incredulously, "But they're a church, how can they do that! Don't they know better?" Things have changed since we only had small print Bibles, hymnals, and typewriters. Accessible materials can now easily be bought or created, but are not available in many churches.

Imagine being invited to a church for a service, or a youth group, or a special meeting to watch an introduction-to-Christianity DVD. Along with warm greetings you receive a bulletin or other material with print too small to read. You sit facing a large screen with small words, poor color contrast, and distracting movement. You may try to listen to singing and preaching, or teaching, but find it is unclear because of distracting background noise. Books around you, a Bible, a hymnal, and other song books, like the text on the handouts or video are all in small print. Some may be incorrectly labeled as large print, even though standard large print books exist, are easy to find, and are not expensive. English is your native language. Anywhere else, you and millions of people with hidden, unobvious impairments such as, not being able to see, hear, or read well, expect, or hope for reasonable accommodations, except in many churches where they could easily be provided. Even some house churches and home groups misuse technology.

We might call this evangelization and think people who decide not to return after their first visits are just resistant to the Gospel or maybe they did not like our style. This could also be called disability discrimination, but many in the church do not know what that means. After all, we treat everybody the same. It seems that we have "some" knowledge of what it means to be human and how to use computers, but not enough.

* * * * * * * *

We may also be using our sound systems inaccessibly. The story is often told of a man who heard a TV announcer describe every detail of a sporting event, but did not hear his wife call him for dinner. This is not about inattentiveness. Men typically hear a low-pitched voice better than a high-pitched one. Consider this; the praise band has a female soprano and a male tenor with a high-pitched voice leading worship. During the service someone indicates they cannot hear the words so the volume is raised. Now everyone is suffering with the painfully loud volume that instead of helping, harms our ears. We lament that more men are not in church, or we think older folks dislike our contemporary music, and wonder why church attendance is declining. All the while, a synthesizer or guitar capo can lower the pitch, and computers can transpose and print the music in a lower key. Instead, we raise the volume and buy more powerful sound equipment.

Computers, sound systems, smart phones, the Internet, and digital cameras are all entrenched in churches today and for good reasons. Yet, even before those were popular, laws were enacted requiring access for people with impairments whenever reasonably possible, even in church. We understand how stairs and rugs are a barrier to people with mobility impairments who use wheelchairs or walkers, but an accessible entranceway, elevator, or lavatory are just the beginning. Barriers to seeing, hearing, or reading, which are less obvious block the life-altering messages we aim to share in church. We learned to evangelize in cities where most people live, but we may not be mindful of people living in the land of disability. These are people we are excluding from church by our habitual misuse of ICT.

* * * * * * * *

I attended a large city high school where only two thirds of the 1,200 in my class graduated. We all might wonder how our classmates fared. I managed to get through, even with low vision, because I was and still am an avid reader as are many who eagerly wait for the latest books to come out in a large print, audio, Braille, eBook, or Online version. After high school I went to a secular college, then to a Bible college and later pursued further graduate studies.

One high school friend of mine who could not read well was a gifted artist. He attended art school and is a commercial artist. Another friend who did not read well went to a trade school and earned a good living. Another friend who was not good in reading or math owns his own successful business. All three graduated in the lower half of our class. In many churches they would not feel welcome because they would find it difficult to read the Bibles, hymnals, or other song books, or projected song lyrics, messages and outlines, announcements, bulletins or paper handouts.

Many people have difficulty reading* a typical small print Bible and find it difficult to participate in church. Not being able to see, hear, or read well are "hidden" impairments. People learn to keep it a secret, to get along without saying anything about the choices others make which exclude them, such as when a church presents inaccessible, print or projected text, or sound. People learn to "get by," say nothing, or not attend.

* * * * * * * *

ᶳ   or reading is not only caused by a learning
        ᴗility or by the lack of instruction. By the 5th
ᵧrade, essential reading skills such as, tracking, scanning, and fluency can be adversely affected by the smaller fonts and increased concentration of letters, words, and lines on a page. This makes it easy to lose your place. These are not the only reasons for a limited reading ability, but they play a part, a part which we have some control over. Song lyrics, or bulletins and handouts, scriptures, outlines, or messages do not have to be printed or projected in small font (small text or letters).

* * * * * * * *

Our technology may seem new, but this issue is not. Seventy-five years ago, C. S. Lewis, in The Screwtape Letters (#2), revealed the Senior Tempter's glee that, "One of our great allies at present is the Church itself." It greets those who enter with a liturgy they cannot understand and religious lyrics in very small print. Today we can use our church's information and communication technology to help clarify instead of obstruct our messages. We did not always have that choice.

Fifty years ago we learned, "the medium is the message." The inclusive use of technology in churches sends a message that God is accessible to those of us who are able to see, hear, and read well and is also accessible to our neighbors who are less able. Marshall McLuhan also noted, "we shape our tools and then our tools shape us." If we allow the Bible to shape us first, we will be trustworthy, living examples of the Gospel.

Consider what Christian workers go through. Pastors, evangelists, missionaries, school, college, and seminary teachers and church secretaries, are all immersed in a very competitive culture where efficient, new technology is highly valued. During an ordinary day, they may rarely or never relate to anyone who cannot see, hear, or read well. They may simplify sermons or lectures, or give one-on-one help, or tutor some, or pray for people, or engage in charitable activities. In their daily work context however, to habitually produce large print or make other modification for youth or adults is rare. Small print is normal to them and to those they typically interact with. They may learn they are required to be inclusive by law, or by policy, or by a professional code of ethics they signed, but they never need to do it, so they may never think about it, or learn how to do it.

New technology for preparing and presenting lessons or sermons can be helpful (Anderson, 2013), but accessibility is essential. Many of us still distribute small print and project hard to see outlines, scriptures, and song lyrics. Industry is tossing must-have, new programs and devices at us. Attending to more things feels overwhelming and being told we are doing something wrong is unwelcome despite instruction being the path of life and intelligence (Prov. 10:17 and 15:32). We prefer suggestions of better ways to do things, rather than hearing about God's commandments or the secular legal requirements we violate.

Some pastors and teachers do have wider experience and produce large print versions of handouts (16 to 72 point font) and set-up larger projected material. Some churches provide large print song lyrics in folders or three-ring binders. One Bible school teacher I know shaved off his mustache so a student who cannot hear, but who can read lips was able to participate in class. Many however, rarely consider access for people with sensory impairments or mistakenly think nothing can be done in church or on a mission field for those who cannot see, hear, or read well, or they may think something will be done by other volunteers or specialists after a request for help.

Of course, not every church owns or misuses ICT. In addition, some systems are simply too small for an auditorium and some churches must weigh priorities and resources because a hearing assistance system costs as much as a computer system. Also, in some churches pastors do all the work. They manage and maintain the facility. They pray, study, preach, teach, plan for growth, including learning new technology, and care for the members. This can be a huge workload, although it is not unlike entrepreneurs in other fields. Even so, it is possible to learn to use technology in inclusive ways all the time and to begin doing it at a reasonable pace.

* * * * * * * *

This is not an advertisement for new and better computer equipment. This book is based on Biblical reasons for being inclusive and on some hidden human variations which effect many of us. Several years ago I declined a request from the editor of an international encyclopedia on disability to contribute an article on the technology used by people who had low vision. This book is not as comprehensive as that encyclopedia and neither is it about special disability technology. Rather, many of the chapters are short lists with only brief explanations and my concern, now as then, is with the technology most churches already own which, if used correctly, could include more of us.

* * * * * * * *

The vast majority of people who have a vision impairment do see, they just do not see well even with glasses or contacts. People with only a basic literacy level can read somewhat. Most people who have a hearing impairment do hear, but not well. How we use ICT affects everyone, but people with these or other impairments may be blocked from participating fully or at all.

The incidence of disability found in America, Britain, Canada, and the world are given in chapter nine. By way of introduction, consider the potential impact of turning communication barriers into bridges. Church attendance estimates (2004-13) indicate 39% of the adults in America, 12% in Britain, and 20% in Canada, say they attend church weekly. What if that same percentage of people with impairments could attend church?

If 39% of the people who cannot see well in America were welcomed in all churches by the miracle of accessible technology, perhaps 23 people who cannot see well could participate in each one of the 350,000 churches in America. If by the same miracle, 39% of those who do not read well, and 39% of those who cannot hear well were also included, each church in America could include 112 of those who have a vision, hearing, or reading impairment. Some attend now.

Some people may have more than one of these three limitations and there are people with other conditions affected by how we use ICT. We cannot assume 39% would attend an accessible church. Not everyone it could help would attend a church for various reasons. We can however, say this adds up to almost forty million reasons to use our technology in more accessible ways.

If all churches became accessible in the United Kingdom and 12% of those who had a vision impairment attended, each of Britain's 50,000 churches might impact 5 people who do not see well. If in addition, 12% who do not hear well, and 12% who do not read well were included, each church might welcome 41 people. This suggests over two million reasons to be accessible.

If all of Canada's 30,000 churches became accessible and 20% of those who had a vision impairment attended church, each church might impact 6 people who do not see well. If in addition, 20% of those who do not hear well and 20% of those who do not read well attended church, each church in Canada could welcome 172 people who might otherwise be kept out. Again, there is overlap. Rather than 172 people per church, this number is better described as more than 5 million reasons for churches in Canada to use ICT in accessible ways.

People usually do not reveal hidden impairments. The goal is that we cultivate the Holy Spirit fruit of self-control when using technology so as to stop harming people by unnecessarily excluding them from church. To do this, we need to know the "why" and the "how" of using the technology we already own, but in inclusive ways that have been required for more than 30 years. The standards presented in Chapter 10 are not new. We need to be mindful of the need to use them all the time.

Inclusive use of ICT is not just a list of laws, demographics, standards or techniques. Neglecting to explain why we do this is like celebrating Christmas or Easter without mentioning the birth, death, and resurrection of Jesus Christ. The powerlessness and the scandal of the cross are the power of God (1 Cor. 1:23-28; Gal. 5:11). How we use technology reflects the Gospel. The Word was made flesh–God became accessible to us. Christ was crucified for us–God was disabled, then died for us, was buried, and rose again the third day. There is no resurrection without the cross and no crucifixion without impairment and disability. This is central to the Gospel.

* * * * * * * *

The chapters in this book repeat and reinforce that theme. Barriers in church begin by ignoring impairments and disability in Scripture. Therefore, chapters two and three offer a disability perspective from the Bible. The two chapters after those present definitions and some laws prohibiting disability discrimination. Disability is not the result of a person's impairment alone. Disability arises when we fail to provide Biblical, reasonable, legally permitted and required access.

The following two chapters look at how a church's culture can make it difficult to change, even though restrictions on the way technology is used is common and any style of church can become accessible. Chapter eight mentions some common, often repeated reasons given for the inaccessible use of ICT, and chapter nine presents demographic information on disability.

Chapter ten and eleven offer more than 80 standards for how and why to use computers to create accessible print, screen projection, a Liquid Crystal Display (LCD), the inclusive use of sound systems, the Internet, e-mail, and social media. The focus is on what is easy to do for little or no cost. Something easily done once can be arduous after 500 or 1,000 times, so a little at a time may be the way to begin.

Chapter twelve, the reference section, contains more than 70 citations. There are more than 140 scripture references in the book. Some of the experiences and training that went into my writing this book are mentioned in the last section, "About the Author," along with an e-mail address for comments and questions, which I welcome and to which I will try to respond.

* * * * * * * *

A limited, one-size-fits-all model for setting up common spaces and activities is an extreme that calls for the least effort to satisfy a two-thirds majority. As a result of this aim, many people miss the worship, Bible reading, teaching and preaching, and other activities. They are not staying home from church by choice. We are actively blocking people's participation in church. Let us instead create better access all the time for those with sensory impairments or limited reading ability. Living the Great Commission and Commandments by reaching more people, with our computers and sound systems, to help them reach out to God, does not require greater effort or an extreme of trying to satisfy everyone's unique needs. There is a balance to our being an approved worker who correctly handles the Word of Truth so that more people will understand it.

Proverbs 3:27 offers a wiser choice, which is, do what is in your power to do. We can easily and inexpensively use information and communication technology to bridge the gap between the Gospel message God has entrusted us with and the millions of people who cannot see, hear, or read well. A better approach is to aim for 95% of the population to begin with. Some will still be left out, so individual help will still be needed when it is available, but the inclusive use of technology by

(1) removing barriers, (2) applying universal design that people with impairments can use, and (3) providing reasonable accommodation, called adjustments in Britain is not difficult, nor is it expensive or extremism. It is wisdom.

A good way to begin is with prayer. To borrow from the Apostle Paul in Ephesians 3:14-21, "Father God, I pray you call all who you desire, to read this book, to understand it and to act on it and that you would accomplish more than we can imagine through it for your glory. Amen."

A child picked a Trillium for his teacher in an area where picking flowers is illegal. She thanked him for his kindness, but asked the youngster not to do it again since it was illegal to pick them. We can thank those who helped us use computer technology and gently teach them why it was wrong and point to better ways, being mindful that habits can be hard to change.

"With all humility and gentleness, with patience, . . . making every effort to maintain the unity of the Spirit in the bond of peace" (Eph. 4:2-3).

2. A Biblical Perspective

As a young teenager, when I believed in Jesus Christ as my Savior and was baptized, my church gave me a Bible with typical print that was too small for me to read even with glasses. They probably did not know large print existed and neither did I. It was twelve years later, by the grace of the Lord, that I discovered a larger print Bible and became an avid Bible reader. I pray you bear with me as I share some of what I read since that discovery and what I think it means.

Hebrews 12:13, Make level paths for your feet, so that the lame may not be disabled, but rather healed. (NIV).

It is from the Bible that we learn people may have an impairment, but their interactions with certain environments are what create disability. Today we have control over the environments we create with our information and communication technology. When we consider impairments and a disability perspective in the Bible we can easily learn why and how to use our tools to set-up church to include instead of exclude, to heal instead of harm people.

Romans 13:10, Love does no wrong to a neighbor.

The Royal Law of Love includes not excluding our neighbors from the message of eternal life or Christian fellowship. However, changing our habitual misuse of computers so as to include those who can see, read, or hear somewhat, but not well, is not a separate >disability< ministry. It is our routine responsibility. The Bible teaches us not to harm or disable people in the first place.

Luke 6:32, If you love those who love you, what credit is that to you? Even sinners love those who love them.

Most groups, Christian or not, will take care of their own when there is a need. Most churches will help someone who they know needs help or who requests help, but people with a hidden impairment are not likely to ask for help. They may know from experience how difficult it is to ask. They live in the same entertainment driven, utilitarian society so know inaccessible churches are intentionally excluding them. They participate less or not at all and leave. A more excellent way is available to us with our computers. We can begin by setting-up church to be more accessible.

We can set-up our communication technology to regularly reach more of the 1½ billion people of all ages at home and abroad who do not see, hear, or read well. For many of these people it is just obvious when a church is not accessible. They see the signs, print notices, books, bulletins, name tags, screens, or try to listen to a sound system that does not have a hearing loop or is too loud and they know whether or not a sanctuary is accessible. Having our messages accessible to the general public, not just "us 4 and no more," helps fulfill the Church's mission.

Proverbs 3:27, Do not withhold good from those to whom it is due, when it is in your power to do it.

Not all churches have the resources to reach everyone and some people are not easy to reach. Today we have the power to reach many more people by how we use our information and communication technology tools. We can remove barriers use universal design and always have alternatives available. The gospel is the power of God to salvation (Ro. 16:1), someone's idea of attractive design layout or music is not.

Accessible use of communication technology is essential for participation in the Word, worship, fellowship, and service. To paraphrase the Apostle Paul, >if technology makes a message unintelligible to some people, how will they know what is being said? Will they not say you are out of your minds< (from 1 Cor. 14:9, 23). Can we have the mind of Christ? The Church's mission is to communicate the Gospel in word and in deed. By misusing our information and communication technology we fail at both with actions that belie our words. Let us instead set up our churches so that more people can see, hear, understand, believe, respond to, and follow Jesus Christ.

Matthew 11: 28-30, Come to me, all you who are weary and burdened, and I will give you rest. Take my yoke upon you and learn from me, for I am gentle and humble in heart, and you will find rest for your souls. For my yoke is easy and my burden is light. This call is not just to the able.

Focusing on people with disabilities found in the Bible and not treating them as if they were metaphors or figurative examples are first steps to our including them in church. We may have traditionally classified them as the "marginalized," who received prayers or charity, those who were once called cripples, or handicapped, that is, beggars with cap-in-hand, or considered them examples of "the weak," which, with a little stretch, could include anybody's or everybody's limits. Instead, we can consider at face value, a disability point of view in the Bible and learn from some obvious, but perhaps overlooked details.

John the Baptist asked Jesus if he were the Christ. Jesus did not say yea or nay. He said, look at how I **relate** to people who do not see, hear, or walk well, or those who are lepers, poor, or dead and decide yourself (Matt. 11:3-5). God heals people from afar by his word and his will. He also came in person and spoke to, listened to, and touched people with severe impairments. He helped people see and hear and understand the Word of God. This is the Christ.

This passage shows that the relationships Jesus had with individuals who had a severe, long term impairment are defining characteristics of the Messiah. From a majority, able-bodied point of view, the focus tends to be on fulfilled prophecy, compassion, and power to heal, which misses the perspective of the people he helped.

We easily overlook their viewpoint if we have no relationship with anyone who has a long term impairment. Prophecy, compassion, and power to heal are important concerns, but a person who is blind who is healed CAN SEE! We can do this somewhat by how we use our ICT. The majority point of reference is not the only way to interpret scripture. Greater empathy with people is possible by how we preach and teach about impairments and disability found in the Bible.

The church often considers reaching people it itself disables the work of volunteers. A meeting between Francis of Assis and Pope Innocent the 3rd was an early entrenchment of the idea that following Christ's commands and example was only for volunteers and not for the church itself. Francis did not ask the Church to change. He only asked that he and his friends be allowed to help the poor and share the Gospel with them. That idea was so radical the Pope had to allow them a special "privilege." Up to and including the Reformation, those who told the Church to change were called heretics. The same is true today. Many churches which disable people by misusing their technology refuse to change. They believe "volunteers" will reach those the church itself unnecessarily excludes.

Next, consider another view of Luke 10: 25-37. To incapacitate means to deprive of ability. It means to disable. In many Bibles, the Parable of the Incapacitated Traveler is labeled the Parable of the Good Samaritan and used to encourage doing good deeds and to discourage racial or ethnic prejudice. Like the lawyer in this passage, many seek to justify themselves. Good deeds are encouraged in scripture, so that interpretation is understandable; "we are ... created in Christ Jesus for good works" (Eph. 2:8-10). Good works are an evidence of salvation; "So faith by itself, if it has no works, is dead" (James 2:14-17). A reasonable interpretation of this parable, however, requires our understanding its context beyond racial, ethnic, or national prejudice, and beyond doing good deeds, in order to see a disability perspective. This parable goes beyond good deeds and shows the need to self-evaluate and confess to whom we are, or are not being a neighbor. Then, it shows the need to demolish man-made barriers to God's love, which Jesus did, which is required for eternal life and for being an accessible church.

This parable helps demonstrate by the title it is typically given the tendency to ignore a person with a disability in the Bible in favor of characters we prefer to, or can more easily relate to. A disability perspective here instead shows the traveler's attitude when incapacitated and relying on a hard-to-believe-where-it's-coming-from-help, which is a faith worth emulating.

The lawyer's first question, the main question, concerns and always does concern eternal life. His second question, "who is my neighbor?" is a barrier to eternal life which prompted a rebuke from the Lord then and now. God's command to love your neighbor (Mark 12:31) is not fulfilled by our local church in our local area by pointing to our missions outreach, which Charles Dickens (1853) referred to as telescopic philanthropy.

Edersheim (1896) noted that in this parable, Jesus turned the lawyer's question "Who is my neighbor?" into "To whom am I being a neighbor?" He has us ask, "What am I doing?" before asking "What should I do?" that is, self-evaluation and conviction before action. If instead of trying to trap Jesus, the lawyer admitted he could not fulfill the law and needed, just like the incapacitated traveler, a hard-to-believe-where-it's-coming-from

Savior, even one from Nazareth, his first question, "What must I do to inherit eternal life," could be answered. The parable's purpose is to rebuke the all too common Rabbinical wrangling over to whom one must be a neighbor. Today we wrangle over or just ignore our neighbors who do not see, hear or read our ICT well.

Kistemaker (1986) also sees that Christ in this parable is revealing God's intent and meaning to "love your neighbor as yourself" by demolishing man-made barriers that obstruct God's law, in this case, the Judaistic rules of whom a Jew legally could or could not help. Today, misuses of our information and communication technology in churches and outreach ministries are barriers that block the Gospel and need to be demolished. We need more than good deeds. We must recognize and demolish self-gratifying technology barriers created to satisfy vain aesthetic standards.

In addition, there is a crucial perspective in the parable of an Incapacitated Traveler, in need of help and receiving a hard-to-believe-where-it's-coming-from help. We all are unable to save ourselves, but are often unwilling to admit it.

Instead we look for a good deed to do so we can be righteous by our own will and strength. We cannot earn salvation; ". . .we hold that one is justified by faith apart from works of the law," (Ro. 3:28) and ". . .we know that a person is not justified by works of the law but through faith in Jesus Christ" (Gal 2:16). The parable's salvation-by-grace perspective may be lost if the focus, like its extra-Biblical title, is the Samaritan's good deeds instead of Jesus' rebuke of our making barriers to fulfilling God's law.

In most churches, some people do good deeds like the Samaritan and some do not, like the priest and Levite. Emulating the Samaritan's compassion is certainly commendable, but Christ specifically focused on people with impairments and he gave us a new law. The old law, to love your neighbor as yourself, impossible as that may be, is still easier and not as sacrificial as Jesus' command to love one another as he loved us (John 13:34, 15:12). We need to examine ourselves and the barriers we build, and admit our need for grace to love others sacrificially, as Jesus loves, not merely as we love ourselves or as we love those who are like us.

Another barrier to becoming an accessible church is to reject the application of scriptures. Schools that use textbooks that deny the Bible applies today undermine the moral fabric of our nations. In their book on ethics, Feinberg and Feinberg (1993) assert that the Bible's command to love our neighbor is general enough to apply to us, but putting up rails on a porch (Deut. 22:8) is specific to a particular culture and not relevant today. When everything is reduced to one word, "love," we miss the specifics of love's outworking and fail to learn how love works. Evidently, those authors never applied for a permit to build a porch.

Rails on porches with precise specifications are still required today because "love does no wrong to a neighbor" (Romans 13:10). The general principle to build our private and our shared environment so as to avoid harming our neighbor, as well as the many other examples of the application of this principle found in the Bible, remain in force today as moral guides and as secular legal requirements. For many people these are just common sense, but ignoring this is a crime, even if denying its Biblical roots and applicability in a book on ethics published by an evangelical publisher, is not.

Laws legislate morality. Government and laws are gifts given to us by God for the protection of us all (Ro. 13:1). Long before our secular laws, the Bible declared our responsibility for what we build. Today it is both law and common sense to put up safety rails on a porch (Deut. 22:8) and not to dig holes and leave them unguarded (Ex. 21:33-34) and not to lead people astray and not to leave things where people who do not see them might trip over them (Deut. 27:18; Lev. 19:14). Projecting or printing a message clearly for people with low vision is a specific application of these God given laws and the general principle not to harm others. Denying these laws ignores God's opposition to unjust discrimination (Lev. 19:15; Pro. 16:8; Acts 6:1; James. 2:1-10). "O you who love the LORD, hate evil!" (Ps. 97:10).

Many places, not only churches, have a policy of being prepared to help those with sensory loss. but they wait until asked for accommodation or adjustments before making things accessible. Our being willing to provide access after a need is know or a request is made may sound like adequate consideration, but it often does not work. People are reluctant to ask for help and this is not new. We think the squeaky wheel gets the grease, but all four Gospels show people were antagonistic toward those who did not see well who requested help. These next examples illustrate how the perspective of people who have impairments in the Bible can help us avoid creating "We'll help if you request it" barriers.

Two men who were blind shouted at Jesus for help and the crowd following Jesus rebuked them and told them to be quiet (Matt. 20:31). When people who were blind came to Jesus in the temple the leaders were indignant (Matt. 21:15). When Bartimaeus, who was blind, cried out to Jesus for help, many followers of Jesus rebuked him and told him to be quiet (Mk. 10:48). When a man cried out for help seeing, those who led the way rebuked him and told him to be quiet (Luke 18:39). Religious leaders insulted the man born blind, whom Jesus healed, and threw him out of the synagogue (John 9:28 & 34). This was pre-crucifixion, pre-resurrection, and pre-Pentecost, but is still a valid disability perspective for today.

People with impaired vision or with any severe impairment are still being discouraged from seeking help by the responses and reactions of Christian followers and leaders. In many ways, the response to a request for help today is still, "Be silent, sit down," or worse. A hostile response to asking for help is still experienced. Potok (2002) noted that people with disabilities are expected to act in a manner that appears to be docile and unprovocative, and undemanding, but just requesting legally required help violates that social obligation. In a study of people with various disabilities other than a vision impairment, Albrecht and Devlieger (1999) found many said requests for access were seen as making waves that led to their being ostracized which was even worse than not having access. They would not ask for accommodation because their experience taught them asking for help lowers their quality of life even more than not being able to participate. The offer of "help if requested" does not work.

This avoidance of help-seeking is not due to feeling embarrassment or to a dislike of being treated like a child or a charity case. People in churches may excel in charity, but also obscure or even object to the difference between charity and justice. The avoidance is due to the active hostility expressed toward those who ask for access to what everyone else has. It is often due to being told by words and by deeds that accommodations or adjustments are odd, or extra, or are being done with great effort, or as an inconvenient favor. Accommodations are integral to one's life and to the lives of many others, and are required by laws in most places.

Another reason not to ask for help is, asking often does not help (Frank, 2000, Johnson and Frank, 2004; Frank and Bellini, 2005). Asking is futile almost half the time and can also be frustrating, difficult, and even dangerous. Frank (2006) found 43% of 311 requests for reasonable accommodation by 151 people with limited vision in America were not fulfilled. The incidence of unfulfilled requests ranged from, 60% of requests for accommodation not fulfilled by government entities, 48% not fulfilled by businesses such as stores, banks, or medical facilities, 37% were not fulfilled when made to employers, and 33% of requests for access in schools were not fulfilled.

Furthermore, the help that is available may be inconsistent and unreliable, especially when under-trained volunteers are part of the process. Accommodation may be offered, but not really exist, or be hidden, or be faulty and not function well, or arrive too late. In addition, many people do not know what to ask for, or who to ask, or how to ask. For many, the burdens are not worth the limited gain. It may be better not to ask, and not to participate in any activity where access is not obvious or readily available–including church. Just as safe and unsafe areas become well known; so too, places that are accessible or are inaccessible become well known.

A disability perspective in the Bible is the view by people with impairments who face disability and the self-examination required by others. Such a perspective helps enrich our view of Christ (Christology), and salvation (Soteriology), Church outreach (Missiology) and Bible Ethics. An appropriate ending to this partial look at a disability perspective in the Bible is what the Word of God says about the end of time, the final judgment (Eschatology). The end comes after the Gospel is preached to all (Matt. 24:14). That "all" certainly includes people with impairments (Luke. 14:13). If we exclude them, when comes the end?

In what may be the first systematic theology on disability, Yong (2010) notes that Jesus' description of the final division between the sheep and the goats indicates the values by which the world will be judged (Matt 25:31-46). How we treat the least members of Jesus' family determines our eternal destination. Matthew 7:1 and Romans 2:1 note it is not up to us to judge each other. There is one judge of us all who disciplines us for our own good, and yet, if we judge ourselves we need not be judged along with the world (1 Cor. 11:31). Considering God's view can lead to self-examination which can lead to conviction and confession, then to repentance, and forgiveness, which can lead to reconciliation, renewal, revival and eternal life.

From these often-read, easy to find, specific commands, stories and parables we get a brief look at a Biblical disability/impairment perspective.

1) The Christ we follow is identified by his helping people who did not see or hear well or who did not understand the Word of God.

2) He demolished man-made barriers to our loving our neighbors, but especially those barriers harming people who are disabled.

3) We are instructed in the Bible to insure that we do not harm, or allow harm to others by what we build or create.

4) Both leaders and followers of Christ resisted requests for help from people with impairments, which teaches them not to ask for help.

5) How we treat the least of Jesus' family will be how we are judged by him.

Therefore, the decades long silence about our misuse of technology must be broken. We could enable instead of disable by printing and projecting clearer, larger text, with better color contrast, and without designs, pictures, or dizzying movement. We could use larger print signs indoors as well as outside and add some descriptive print and audio captioning to video. It is also possible to build accessible Web sites and to install doors that a person who is frail or uses a wheelchair or walker could open. More suggestions are offered in chapters 10 and 11.

Rather than evoking mere physical excitement with the power of electricity, we could lower the pitch and the volume and the electronic effects of our sound systems so people just home from the hospital after another round of chemotherapy, a heart attack, or a stroke could endure our worship and attend church again. We could install a hearing assistance system to our sound systems for those with hearing loss and advocate for Sunday bus service. With these in place we would not need to rely only on volunteers

God's response to the Apostle Paul's three requests for his own healing was, "my grace is sufficient for you, for my power is made perfect in weakness" (2 Cor. 12:9). "God chose what is foolish in the world to shame the wise; God chose what is weak in the world to shame the strong" (1 Cor. 1:27). God's grace and power can still be demonstrated through the Church.

Cursed be anyone who misleads a blind person on the road (Deut.27:18).

You shall not revile the deaf or put a stumbling block before the blind; (Lev. 19:14).

> For God has done what the law, weakened by the flesh, could not do: by sending his own Son in the likeness of sinful flesh, and to deal with sin, he condemned sin in the flesh, so that the just requirement of the law might be fulfilled in us, who walk not according to the flesh but according to the Spirit (Romans. 8:3-4).

Give instruction to the wise, and they will become wiser still; teach the righteous and they will gain in learning. (Proverbs. 9:9).

> The seeds of the simple Gospel sprout into a relationship with God, leading to changes in our relationship with others and with our world, and with our use of technology.
>
> Those seeds can grow if we learn the gospel is not just about us or those like us.

The next chapter presents applied models of accessibility–planned before the foundation of the world. It was not left up to those who needed it.

3. Biblical Analogies

In the previous chapter, several scriptures were highlighted from a perspective of people who had severe impairments. Hundreds of other scriptures mention people with impairments and many more can apply to the topic when viewed as analogies, that is, they contain similarities suggesting a comparison between a scripture's principal focus and some aspect of disability. Applying analogies can be difficult. Application is referred to as going from preaching to meddling. It may help if we first compare two kinds of fruit of the Spirit; love and self-control.

Over the past three decades a chasm was created by our misuse of technology. A bridge across that divide must be built from both sides. Those who have been, and continue to be disabled and excluded by the churches' ICT must cultivate the fruit of love (1 Tim. 1:5; 1 Jn. 4:7-8), such as, turning the other cheek and engaging with the local church. The disabling church must embrace a greater, but Biblical range of human variation and learn new ways of working, which are possible by cultivating the spiritual fruit of self-control (Gal. 5:23; 2 Peter 1:6).

A church should be accessible whether or not anyone with an impairment is disabled there. The steps for this are for us to exercise self-control. We also need to not keep it a secret, but publicly let it be known that we aim to be accessible.

To understand the differences between good deeds motivated by love and the practice of reasonable self-control, consider that when we give someone a ride in a car we see the person helped. In contrast, when we fulfill the law by driving safely, we usually do not see the people we did not hurt or the accidents we avoided. The difference is also highlighted by Paul's concern that he not be disqualified (1 Cor. 9:27).

Suppose Christians, in their home country or abroad, on their way to love motivated ministries, drove at twice the speed limit and ran people over. They would be disqualified. Misuse of technology does matter. Ophthalmologists perform surgeries in mission hospitals around the world. Members of Lions clubs collect old glasses and with the help of optometrists they too help people see. Old hearing aids can also be recycled. In some churches, volunteers teach reading, or they teach English, or computer word processing, or create large print and Braille Bibles to give away.

Those uses of technology demonstrate the fruit of love, but the ministry would be disqualified if proper self-control to do those things correctly was not exercised. Just being fast or occasionally being accessible is not acceptable.

Paying an electric bill, obeying traffic laws, or using ICT inclusively instead of exclusively are examples of behaviors needed when using technology. They are not good works or acts of charity. They may help people and include the fruit of love, but we usually refer to them as required personal responsibility, discipline, and self-control even if those terms are not popular. They are not artistic licence or attractions.

* * * * * * * *

The following four analogies are examples of God's love and self-control and his inclusive communication. The greatest ACCESSIBILITY story ever told is found in John 1:14, "And the Word became flesh and lived among us, and we have seen his glory, the glory as of a father's only son, full of grace and truth." and in Eph. 2:17-18, "So he came and proclaimed peace to you who were far off and peace to those who were near. For through him both of us have ACCESS in one Spirit to the Father." We are redeemed because God revealed himself in the ACCESSIBLE form of his Son, "through whom we have obtained ACCESS" (Ro. 5:2). We have ACCESS to God because Christ died for our sins, was buried, and was raised on the third day (1 Cor. 15:3-4). Do we deny it to others by the way we use technology?

The most astonishing example of DEMOLISHED BARRIERS to God's love is noted in Mark 15:38, "And the curtain of the temple was torn in two, from top to bottom," and Eph. 2:13-14, "now in Christ Jesus you who once were far off have been brought near by the blood of Christ. For he is our peace; in his flesh he has made both groups into one and has broken down the dividing wall, that is, the hostility between us." He is still breaking down and demolishing the unnecessary barriers we build, even the walls of technology and our overwhelming busyness with everything else that needs to be done. This is not a call for extra work or to do more good deeds. We need to relearn what we already do, but learn to do it correctly, accessibly. This is also called repentance, which is recognizing and admitting we are doing something wrong, stopping and turning from it, and going in another direction. It allows forgiveness and reconciliation, which is renewal and revival.

The most user-friendly **UNIVERSAL DESIGN** is found in Matt. 11:28 (NKJV), "Come to me all you who labor and are heavy laden and I will give you rest" and John 7:37, "Let anyone who is thirsty come to me," and Rev. 22:17, "And whosoever will, let them take of the water of life freely." All who will come can come and partake, not only those who are able-bodied. Not everyone will come, but those whom the Father calls will come. We are called to help prepare the way for them–a way without stumbling stones (Isaiah 62:10).

By far, the most valuable **ACCOMMODATION OR ADJUSTMENT** ever provided to helpless, fallen humanity is described in Eph. 2:8, "For by **GRACE** you have been saved through faith, and this is not your own doing; it is the gift of God." We neither paid for, worked for, nor deserve this accommodation that allows us to participate in God's Kingdom and to share the good news with others, "Freely you have received, freely give" (Matt. 10:8 NKJV). We are called to be imitators of God (Eph. 5:1; 1 Jn. 2:6) who willingly became accessible to us. Being an accessible church sends a message that God is accessible and the message of God's grace to us and through us.

* * * * * * * *

"The hearing ear and the seeing eye, the Lord has made them both." (Pro. 20:12). God's opinion of ingratitude is found in the parable of the man who was forgiven a huge debt he could not pay, but who then did not forgive a very small debt which he was owed (Matt. 18:23-35). Three quarters of the population in America (about 225 million people) use glasses or contacts. We can show our thankfulness for those technology gifts that help so many of us by using our other technologies in church to include people who cannot see well even with glasses or contacts. When we provide large print or accessible computer projection, things which are easy to do and essential for millions of people, we demonstrate we are thankful for our many gifts.

The telephone was invented as an aid for people with hearing loss. We can show gratitude for that beneficial gift of technology by using our sound systems in church to include people who do not hear well. Instead of increasing the volume so it harms people's hearing, we can lower the pitch and add a hearing loop or other hearing system. Human variation impacts us all. What will be our grateful response to God's grace to us?

The problem is not a scarcity of volunteers to do extra work on additional priorities. Our initial work and priorities are wrong. We habitually set up our churches using technologies incorrectly every week. Whenever it is reasonably possible and does not alter the essential nature and purpose of an activity or space, we need to learn to be inclusive to begin with, before we set things up incorrectly. The knowledge is not hidden. It has been ignored or rejected and bad habits are sometimes hard to change.

A Biblical view of disability, whether precisely found in scripture or inferred by analogy, does not omit advocacy. The Bible goes beyond studies of attitudes or social stigma. The Bible calls for advocacy, which is action that produces changes in behavior. "Speak out for those who cannot speak, for the rights of all the destitute. Speak out, judge righteously, defend the rights of the poor and needy." (Pro. 31:8-9).

Timid Christianity is uninspiring and lukewarm. Who will display the courage of John the Baptist and preach about the high incidence of divorce and its financial consequences when a family member experiences a disability (Singleton, 2009)? Who, like Jesus, is overturning the money lenders' tables which blocked Gentile worshipers from the court of the Gentiles in the Temple? Today those tables are piled high with illegally inaccessible notes. Years ago, the United States Treasury Department lost its court case and was ordered to print accessible currency like other countries do, but the U.S. Treasury Department still does not do it (O'Connor, 2013). Our silent voices must become louder than the hawking shouts of the technology marketplaces in church.

* * * * * * * *

If analogies between redemption and including people with impairments in church seem a stretch, consider whether they contradict any other part of the Bible. All have sinned and fall short of the glory of God (Ro. 3:23). No scripture excludes people with impairments from approaching God. Perhaps people with impairments were excused from service in Lev. 21:16-23, from the arduous labor of Temple worship out of kindness. It is likely that the difficulty of the 24/7 shift work of hauling water, wood, ashes and slaying animals, and not their blemishes, was what would profane His sanctuary. God is not a hard taskmaster. Are we cruel because God is kind? He allowed them to partake of the most Holy bread and the Holy things. Do we?

We can include people who have low vision or hearing as leadership, staff, and volunteers in church and Christian ministry. Think of Sunday school or Bible study teachers and helpers trying to read a small print, computer-generated class list through bi– or tri-focals. We can make some or all printed material, including sign-up sheets, in larger text (14-16 point font). We can install a sound system and hearing loop in classrooms.

A large print Bible (see page 132) was essential for me as a pastor and as chaplain of a nursing home, a college, and a hospital. As a counselor at a Christian rescue mission I reformatted the handbook everyone used into larger font. As a house supervisor at an Arc group home and when helping at a Friendship Club I found larger print would help with such practical materials as the directions posted for using a heater, an air conditioner, a hot water unit, and for a list of emergency phone numbers. My training as a therapy aid at a State psychiatric center included distributing medications, a critical area where larger fonts help. Making such simple changes allows people, with and without impairments, to grow together in the grace and knowledge of our Lord and Savior Jesus Christ (2 Peter 3:18).

The hermeneutic method of explaining scripture in this book begins with St. Augustine's prerequisites of charity towards God, neighbor, and self (1 Tim. 1:5,), and not claiming exclusive knowledge (2 Peter 1:20).

A golden rule of interpretation is that when the plain sense of the scripture makes common sense, seek no other sense. Take every word at its primary, usual, meaning, unless the facts of the immediate context, studied in the light of other related passages and fundamental truths, clearly indicate otherwise. Face value, or prima facie evidence, is evidence sufficient to establish a fact or raise a presumption of fact until refuted by other facts.

Secular laws prohibiting disability discrimination are described in the next chapter.

4. Grace and Law

They used technology that kept us out.
Contemporary, ambiance, aesthetics they shout.
But love and I had the wit to win.
We passed some laws that allowed us in.
(Paraphrase of Edwin Markham's
poem, "He Drew a Circle.")

Some laws prohibit disability discrimination, others permit creating access. After graduating college my eyesight was getting worse, but I had no insurance. Reading glasses were expensive because by law only a professional could prescribe them. Then, a friend from church bought a pair for me. Years later, laws were passed that allowed drug stores to sell them over the counter for about $20 a pair instead of $100 or $200 via a professional. Today, that kind of access is suggested for hearing aids too which would allow them to be purchased at lower cost.

Occasionally someone in a church erroneously suggests changing formats of printed or projected material violates copyright law. Quite the contrary, access for people who do not see, hear, or read well is authorized by copyright laws and treaties. Churches, as authorized entities, can create, save, and distribute accessible formats. Again, this book is not offered as legal advice. See the 1996 Chafee Amendment, [17 U.S.C. § 121], at (www.loc.gov/nls/reference/guides/copyright.html). Also see Article 2.c of the signed and ratified, 2016, Marrakesh Treaty to Facilitate Access to Published Works for Persons Who Are Blind, Visually Impaired, or Otherwise Print Disabled, at the World Intellectual Property Organization (wipo.int) site, (www.wipo.int/meetings/en/doc_ details.jsp?doc_id=241683), Learn more about the Marrakesh Treaty at (www.worldblindunion.org).

Secular laws, treaties and definitions were developed thousands of years after similar Biblical principles were given, noted in chapters 2 and 3. One treaty giving an overview of laws permitting or requiring access for people with impairments is the International Classification of Impairments, Disability, Function and Health (ICF), (WHO, 2001) ratified by 191 countries.

The ICF-2 defines impairment and disability and where they takes place.

1. An Impairment is an injury, illness or congenital condition that causes or is likely to cause a loss or difference of physiological or psychological function.

2. A disability is the loss or limitation of opportunities to take part in society on an equal level with others due to negative interactions with social and environmental barriers. Those barriers can be grouped under five headings:
- negative cultural representations
- inflexible organizational policies, procedures and practices
- segregated social provision
- inaccessible information formats
- inaccessible built environment and product design.

Impairment is part of a disabling interaction, but it is not the cause of disability. Disabling barriers in those 5 areas mentioned by the ICF-2 are the cause of disability discrimination and are found in churches. The problem is people do not know about this or do not like being told what to do or not do and they think what they do is right and their right to do. Being required to pay an electric bill is acceptable, but being required to use electricity in accessible ways, and not for just a two thirds majority, is considered unacceptable interference. Voluntary, individual help for the excluded minority if needed is preferred over setting up church to be accessible in the first place. Therefore, laws were enacted requiring reasonable comprehensive systemic changes.

Definition: Disability occurs in the interaction between features of a person and features of society which we create. Access for people with impairments requires removing, where reasonably possible, physical and social barriers. Access is thwarted if churches never learn about the methods and laws, which according to the preamble of the ADA, are intended "to establish a clear and comprehensive prohibition of discrimination on the basis of disability." Churches fail if they try to apply those laws and methods one person at a time instead of comprehensively in all they do.

Discrimination on the basis of disability has been illegal for more than 30 years in the United States under the Americans With Disabilities Act of 1990 (ADA), now the ADA Amendments Act of 2008 (ADAAA, 2008). Since 1995, it has been illegal under Britain's Disability

Discrimination Act (DDA), now part of the Equality Act (EA) of 2010. It has been illegal for more than 30 years under the Canadian Charter of Rights and Freedoms of 1982, and with enforcement provisions, since 2005, under the Accessibility for Ontarians with Disabilities Act (AODA). These or similar laws at local levels apply to churches in all three countries, yet mention of laws against "disability discrimination" is rarely or never seen in Christian media. Secular media is equally silent (Johnson, 2003).

Laws against disability discrimination exist because we create much of our environment and produce the things we use. We can regulate the things we build and manufacture. Creating equal access requires acknowledging responsibility for the consequences of what we create. The debilitating result of poor vision or hearing, or limited reading ability is not caused by an individual's impairment alone. We create barriers. It is not necessary to produce small print, project hard to read messages or misuse sound systems. We could easily and inexpensively be inclusive to begin with as we set up church each week.

The Bible and secular law often, although not always coincide. Laws prohibiting disability discrimination are similar. The laws involve more than can be covered in this brief book. This is only a slight sketch of the legal context and environment of churches for the purpose of introducing the definitions and laws and how they work. This is not offered as legal advice. Sources for further information are provided for each law.

United States: In enacting the Americans With Disabilities Act (ADA), the United States Congress found, people with disabilities face discrimination barriers that keep them from participating in **all** aspects of society including such critical areas as **communication**. The law's purpose is to eliminate those barriers by providing a comprehensive national mandate with clear enforceable standards. The ADA prohibits several types of disability discrimination, including; the failure to build in accessible ways for people who have a seeing, hearing, walking, or other impairment. It is illegal to fail to remove obstacles when it is reasonably possible to do so; or to create inaccessible activities that could reasonably be made inclusive; or to fail to provide in a timely manner reasonable accommodations (called "adjustments" in Britain) that allow participation by people with impairments. Access for individuals and being generally accessible are both required.

Those definitions of disability discrimination are used in local laws and in other countries as well. "Reasonably" or "reasonable," means that the law does not require altering the essential nature of an activity or space and also that adherence to the law should not be too costly or too difficult (ADA, 2009, U.S. Department of Justice, 2012). About 20% of the population has a severe impairment, so using about 20% of our resources for the purpose of accessibility is an estimate of the terms "not too costly" and "not too difficult."

Everyone in the United States is protected by the ADA, but churches lobbied to be exempt from the law due to claiming it violated their freedom as religious organizations. Even so, a church with five or more employees, like any business, may not discriminate against present or potential employees due to their having an impairment. However, if people with an impairment are blocked from participating in church, how could they acquire a desire to work there?

Bickenbach (2000) observed that laws which prohibit disability discrimination are complicated to utilize and enforce. They may be unknown or not well understood. Government relies less on litigation and more on education and mediation to enforce the ADA. For many people, the complaint process in the United States results in "winning" nothing except a too costly right to sue (Moss, Burris, Ullman, Johnsen, and Swanson, 2001). Then, for the very few who do sue, the court system results in an over-whelming number of losses (O'Brien, 2001). Forced compliance or compensation are rare. The ADA amendments of 2008 provided clarification of the law's intent, but to gain that, a compromise, further weakening its enforcement was required. Paul admonished believers not go to court against believers (1 Cor. 6:1-8), but people with impairments have learned for themselves to forgo the ineffectual ADA complaint process (Frank, 2006). Moreover, who would want to participate in a church or in any place where you had to go to court to force your way in?

On the other hand, but also rarely discussed, State and local governments in America regulate churches now. We can be thankful that churches are subject to local health, sanitary, and safety laws, such as laws for fire exits, or the number of people allowed in a building, and safe electrical wiring. Access for people who use canes, walkers, or wheelchairs has improved in churches due to local laws which require accessibility be part of any plans before building, renovating, or remodeling permits are granted. Churches can be fined for noncompliance. Nevertheless, after the electrical wires are connected from the outside, what we do with computers and sound systems inside churches in America is largely unregulated unlike in nonreligious organizations and unlike laws in countries that do not emphasize the separation of church and secular government.

Britain: The Disability Discrimination Act (DDA) of 1995 requires "reasonable adjustments." Service providers must provide access to goods, facilities, services, education, and premises. This includes information and communication. Churches in the United Kingdom (UK) are covered by this law and the later Equality Act of 2010. It is unlawful for service providers to treat people with disabilities less favorably for a reason related to a disability. Service providers may have to make reasonable adjustments in relation to the physical features of their premises in order to overcome physical barriers to access.

The Equality Act of 2010 (EA) combined the DDA with other nondiscrimination laws. It also protects persons associated with a disabled person such as caregivers. For more information, see the free, 19 page, .pdf handbook, Disabled Consumers and the Equality Act 2010, at: www.dls.org.uk/advice/factsheet/consumer_contr act/goods_services_EA/Goods%20and%20Servi ces%20and%20the%20EA.pdf Also see the 26 page book, Enforcing your Rights as a Disabled Consumer, at: www.dls.org.uk/advice/factsheet/ consumer_contract/enforce_rights/Enforce_Your _Rights_As_A_Disabled_Consumer.pdf.

Under the DDA the government provided some legal advice, and in 2000, the Disability Rights Commission (DRC) received additional power to pursue agreements in lieu of enforcement action. The act was strengthened in

2005. Although the Equality and Human Rights Commission has some enforcement powers in certain situations, its limited resources means it cannot handle many cases. Under the EA of 2010 it is still up to the individual to go to a court or tribunal to lodge a complaint. The process is an individual ordeal by a person who has a severe impairment and few or no legal resources, against a large organization with legal resources.

Dickens (2008) notes that different standards for public and private sectors makes the EA less efficient and its emphasis on resolving individual complaints rather than seeking changes in the discriminatory behaviors of organizations makes it less effective. The fact that complainants must try to have their rights enforced in an adversarial tribunal system where procedures are complex and usually slow, and where most unrepresented applicants are at a disadvantage, calls into question the equality of access to justice. Despite changes designed to streamline the EA, Dickens (2014) is not optimistic about its effectiveness.

Canada: The Accessibility for Ontarians with Disabilities Act (AODA) became law in 2005 with full implementation in 2012 and the goal of an accessible Ontario by 2025. It is a continuation of other laws protecting people with impairments, including, the Canadian Charter of Rights and Freedoms of 1982, the Canadian Human Rights Act of 1985, and the Ontario Human Rights Code of 1992. The Ontario Human Rights Code prohibits overt discrimination as well as covert practices that are discriminatory in their effect.

These laws establish the principle of equality of access for persons with disabilities to goods, services, facilities, and employment. The law requires being accessible, not merely providing access one case at a time. It applies to public and private organizations, including churches and educational institutions.

Over twenty years ago, I discussed this issue with scientists and government officials at the Ministry of Public Health in Ottawa, Canada. They knew the scope of the issue and devoted attention to determining how the country should best proceed. Laws were proposed based on various population predictions. The laws and the predictions have come to pass.

The AODA provides legal mechanisms for individual complaints, enforcement and financial penalties, but the Accessibility for Ontarians with Disabilities Act Alliance (AODA Alliance, 2015) reports that the government is not enforcing the AODA in public or in private sectors despite having over 20 million dollars earmarked to enforce it. This is another David versus Goliath story with the twist that David may have a severe impairment or not see, hear, or read well.

The AODA requires that **communication** with a person with a disability take into account a person's disability, and that staff be trained in serving and interacting with people who have disabilities. It requires an accessible feedback process so the public will know the accessibility status of an organization and of a particular request for access. Guidelines for the AODA are available in a 77 page .pdf document, "Accessibility Standards for Customer Service Handbook" (2009) (www.mcss.gov.on.ca/documents/en/mcss/ accessibility/Tools/AO_EmployerHandbook.pdf). Also see the 44 page .pdf booklet "Information and Communications Standards" (www.mcss.gov.on.ca/ documents/en/mcss/accessibility/iasr_guidelines/ Part2_IASR_2012.pdf).

The enforcement of the AODA in Canada is likely as difficult and rare as the enforcement of the Equality Act in Britain and the ADA in America. Local city laws may be more effective. **Laws prohibiting disability discrimination that are often unenforced and ineffectual apart from the good will of the citizens of a country are the context and environment of the Church.**

We may feel no urgency to use information and communication technology in accessible ways because we have little or no fear of civil penalties. Or, we can be inspired to reach for our high calling to be the light of the world by being obviously accessible. That choice would be ours to make, if we knew it existed: if the laws and scriptural reasons the laws exist were preached, taught, and discussed in churches, Christian schools, and in Christian and secular media. Instead there is silence or misinformation and opposition. Biblical and secular laws prohibiting disability discrimination are too often ignored or diminished. Even if we hide from laws, we cannot avoid them, "As for those who persist in sin, rebuke them in the presence of all, so that the rest also may stand in fear" (1 Tim 5:20).

The United Nations: The UN Convention on the Rights of Persons with Disabilities (CRPD) was, adopted in 2006 and signed by 165 countries. It includes similar definitions and many of the same requirements that were mentioned above (see: www.un.org/disabilities/convention/conventionfull.shtml). The UN World Health Organization (WHO) distinguishes between people's impairments and the disability which is caused by their interaction with certain aspects of society. The UN CRPD states that persons with impairments, in all parts of the world, continue to face barriers to their participation as equal members of society and affirms the importance of accessibility, especially to **information** and **communication** systems.

The signers of the CRPD, including Canada and Britain, agree to eliminate disability discrimination and barriers to accessibility, including the refusal to provide accommodation or adjustments, in order to ensure that persons with impairments have equal access. The United States signed the treaty, but has not ratified it, (see the discussion at: www.ncd.gov/publications/2014/07142014/). One reason suggested is that Churches in America are exempt from some parts of the ADA, not so they can discriminate on the basis of disability, but to insure that the Federal government does not support religion. As noted above, local anti-discrimination laws in the U.S. do apply to church.

The CRPD requires the removal of barriers and collection of data concerning the impact of the law on people with disabilities This data can be misleading. Some years ago I evaluated ADA complaint and mediation data which found that thousands of complaints were "successful." They revealed real barriers leading to valid complaints exist. Unfortunately, "successful" mediation did not reveal if barriers were removed or if any people with disabilities were helped by filing complaints. That more important information could easily be collected, but was omitted.

The CRPD mandates equal access to justice, but as mentioned for the ADA, the Equality Act, and for the AODA, utilizing the legal system is often difficult and costly. This is not unusual. Laws protecting anybody, against petty crimes, or even for more serious crime, are often not enforced even if an attempt is made. There are not enough police or judicial resources to address all violations of law within our legal systems. Choices of what to prosecute must be made for every law. When disability discrimination law is ignored, opposed, or has a low priority, equal access for people with disabilities means equally slow, limited, or denied justice.

Secular laws articulate moral principles often found in the Bible. Nonetheless, except for a few very grievous, lucrative, or large demonstration cases, equal access is based mainly on the law-abiding character of a society, that is, its moral fabric – which churches help create. The Church is called to be the light of the world (Matt. 5:14). Even so, opposition to these laws exists in churches as well as from secular sources. A church may be inaccessible, but not know it is their choice or that there are no accessible churches to attend. There is little current data on church use of technology or attendance by people with impairments. It can be unpopular data to collect. Since most people are loath to file a complaint against a church, its disability discrimination may often only be revealed by seeing expenditures and efforts which, instead of being inclusive, only benefit the able-bodied.

A survey some years ago found just under half (49%) of the people with a disability in America attended church at least once a month compared to 59% of those without a disability (Kaye, 1998). More recent estimates suggest four fifths (80%) of those experiencing disability themselves or in their family can no longer find a church to attend (Vander Plaats, 2014; Stumbo, 2014). Anyone can easily verify those estimates, as I have, by contacting churches by phone or in person.

A visit to a dozen churches in one area revealed they all used technology. Three out of 12 provided large print material (by photocopy) and only when it was requested. First, I called during the week, to prepare in advance and then, on Sunday when nothing was obviously available, I asked again from the ushers distributing small print material. One just laughed at me. I also called the main offices of churches and church denominations to ask if they made provision for people with disabilities. A few responses directed me to separate special education services for people with special needs, such as, those with Intellectual and Developmental Disabilities (IDD). But the people I spoke with in those churches had never heard about modifying how technology is used for people with a sensory impairment, or who did not read well and they would not do it.

How we use technology may have contributed to a huge decline in church attendance by people with sensory impairments and their families–who could become any of us. Disability discrimination in the way a church uses ICT has the affect of excommunication and disfellowship of members of the church without even a trial (Ps.42:1-5). This negatively impacts everyone. Without a Christian ideal to attract the young and with less justice in church than in the rest of society, the inaccessible church is unattractive to the able as well as to those it disables.

Part of the research for this book was to read Christian books and articles on the topic. There are many that focus on people with IDD or "special needs," but none mentioned self-control of technology by the able-bodied, or the law. Some stressed attitudes were the greatest barrier, which can be true for some types of impairments. Some tried to create a new theory or theology of disability, or felt details of access were already well known. Few mentioned disability discrimination or its demographic scope.

Some books and articles oppose the laws by suggesting they harm the very people they intended to protect. Bagenstos (2004) exposes that "perverse-results" argument as the attempt to give the appearance of supporting the people who face discrimination, yet opposing the laws protecting them, but "for their own good." A "perverse-results" excuse in churches may take the form of opposing inclusive use of ICT in order to reach others with a certain "aesthetic style." That approach is saying, "we are unable to share the Bible if we abide by the Bible." It denies the power of the Gospel and seeks to continue discrimination. Such opposition is not surprising. Our technology gives unprecedented potential to fulfill the Great Commandments and Commission which is opposed by the enemy of Christ.

These laws are also opposed by charges of, "excessive government intervention in church." Such exaggeration tries to promote fear of the law. Another way the laws are opposed is by referring to the crime of disability discrimination as if it were about manners or hospitality. That spin tries to minimizes the law and diminishes our responsibility. Disability discrimination is a crime and a sin. It is not a question of manners or hospitality. People with impairments may be life-long members of our church who would attend and participate if we made it less painful and more accessible. They are not only strangers to whom we should extend hospitality. Charity, often seen as pity, is not the answer (Shapiro, 1994). Substituting benevolence for justice only offends people who have lived for more than 30 years in countries where technology barriers are illegal.

Valid questions regarding the effectiveness of laws against disability discrimination in church or anywhere else exist. Of greater concern is our churches' silence concerning the law, disability, the range of human variation, and the Biblical basis and mandate for the inclusive use of technology. If we speak up and attend to an issue there is often measurable success.

My apologies if I missed or misinterpreted any book, but I believe, at least with ICT, behaviors are a greater barrier than attitudes and are more readily changed. Laws do help keep us from committing mean and thoughtless acts. Every time I read through the Bible I see people with impairments are valued and included in God's family, church, and Kingdom. If those scriptures are taught and preached attitudes will change. There has always been a theology and a grand theory concerning impairments, disability, and inclusion. It is called Christianity. Therefore, the requirement to know, discuss, and apply the details of accessible use of ICT is needed by churches.

We have the tools to easily include more people. If we demolish man-made barriers, build with universal design, and provide reasonable accommodation and adjustments we can share a message of, "acceptance by God" in a way that says, "we accept you." We can, in humility count others more significant than ourselves, and look not only to our own interests, but also to the interests of others (Phil. 2:3-5).

Soon after this book was announced in Britain, a statement was posted on the Church of England's Website that churches in England are subject to the Disability Discrimination/ Equality Act. In less than a fortnight that notice was replaced by pages of outcry against government intervention in churches. A search of that website revealed that their most comprehensive disability access audit made no reference to how to create accessible websites or even clear print guidelines.

Such fear mongering and exaggeration occur in America too, and also without mention of how to be inclusive. We must demonstrate that we are citizens of a better country, one whose King and kingdom are accessible.

* * * * * * * *

The Bible says, "it is not in man who walks to direct his steps" (Jer. 10:23) and "Let every person be subject to the governing authorities" (Rom. 13:1-7), and "For the Lord's sake accept the authority of every human institution" (1 Peter 2:13-16). Jesus said of a secular law, "if anyone forces you to go one mile, go also the second mile" (Matt 5:41). Laws which prohibit disability discrimination are part of the environment and the context of the church and are also part of the Great Commandments and the Great Commission.

Laws controlling the use of ICT are not unusual. The next chapter explores some ways technology is regulated throughout society—even in churches.

Writing is a technology. We have the Bible due to the accuracy of the scribes following the rules when copying it. We are still called upon to use our tools to correctly handle the Word of God (2 Tim 2:15) in order to share the Word of God with as many people as possible (Matt. 28:18-20).

* * * * * * * *

A pastor tried out his new car on an empty road. Aerial surveillance clocked him at 110 mph (about 190 kph). He got a ticket in the mail. Authorities later found out he was in the country on a work permit and was not a citizen, so he was deported. It took him some time to get back in the country during which time the church was without its shepherd. **The use of technology is restricted and someone is watching**.

* * * * * * * *

At around age 45, many people experience vision changes and need glasses with bi-focal or tri-focal technology. This is a normal aging process and is not usually considered a vision impairment. Reading smaller print is possible with those tools, but the eyes and body may need to be carefully positioned and held steady to read. This can be tiring, even painful over time. Larger text can help.

5. Restrictions On Technology

A few students from a nearby university, some using white canes visited our church and were led to the coffee room where I was the only person to greet and talk with them. Laws against disability discrimination do not require good manners or hospitality. Changing people's hearts or attitudes; welcoming and including people in our feasts of worship and the Word, or even liking a person with a severe impairment is not required by secular laws. Those are the purpose and work of the Holy Spirit and the gospel (Luke 14:13-23). Secular laws only require that we be accessible; that we provide reasonable access in what we build and create, including how we use our technology, whether or not anyone with a severe impairment is present or requests it.

"You shall receive power when the Holy Spirit has come upon you" (Acts 1:8). We also receive power when we use technology, be it a train, a jet plane, a car, a sound system or a computer. We can be faster, louder, smarter, and more powerful with technology and that is hard to restrain. Yet, restricting how we use all types of technology is normal for all of us, even for those who feel they have a vested interest in the "newest thing," and how it is used. Paul's discussion of justice, self-control, and coming judgment, was frightening (Acts 24:25). Today, discussing the restraint of technology is rare, and even alarming to some. Regulating how we use technology is normal in both religious and secular organizations.

Many churches learned how to use technology incorrectly or went along with what volunteers did. Still, it is not written in stone or inviolate and learning how to use new programs and products is going on all the time. At the same time we reach out with charity, good deeds, evangelizing or missions, we cannot neglect our responsibility to carefully use all our latest tools in well-known, inclusive, and nondisabling ways.

Just following the status quo disables people with impairments and excludes many. We can learn to build bridges and not harm and exclude people, but include them instead by changing what we have done wrong for decades with information and communication technology. Admitting fault may be more difficult than learning new tricks and breaking bad habits. It is called confession.

The government repeatedly uses media to remind the public of the various restrictions on technology, such as, giving warnings of the stiff fines for not obeying speed limits, not wearing seatbelts, or not having the required number of passengers in an express lane. Motor vehicles were used for decades before safety features were required by law. The first standards enacted fifty years ago were seatbelts, dual braking systems, padded dashboards, and windshield wipers. Sadly, for years before safety measures were required, the auto industry knew those features would save lives and sold them one at a time to individuals, but refused to apply them universally, claiming the added cost would be unpopular with the public.

Safety issues were ignored until the auto industry and the public were forced by law to act. Thanks to laws restricting the manufacture and use of transportation technology, the automobile death toll in America dropped from 5.3 to 1.1 deaths per million miles traveled (the National Highway Traffic Safety Administration, 2014). How many lives could be saved for eternity if churches fulfilled the intent of the Bible and laws requiring access for people with impairments?

Auto safety laws are not mentioned here because auto accidents may cause disability. The point is our reluctance to restrain how we use technology until forced to do so by law. Recently, we were unwilling to forgo using cell phones to talk or text messages while driving, so hard-to-enforce laws were enacted to protect us from our dangerous use of that technology.

Some colleges prohibit freshman or sophomores from having cars on or even near campus. High insurance rates for younger drivers inhibit some young people from driving until they are older and (hopefully) wiser. Nowadays movies and concerts begin with requests to shut off personal electronic devices because many of us lack self-control in regard to the use of our personal technology. Some people are addicted to cell phone use. Restricting technology is normal. We may not like external controls, but, most agree we need them.

We like technology, but when it causes harm, we control it, or fix it, or change how we use it and provide warnings. Many Americans fiercely defend their right to bear arms, but in most communities it is illegal to discharge a firearm. How we make and use any technology--not just ICT is restricted by law.

When a church ignores this, it harms the people Jesus told us to bring to our feasts. One pastor asserted he would not use his church's technology in inclusive ways until he was paid to–or forced to do it. Perhaps he was thinking of all the work of trying to do it by himself all at once. Church has never been perfect. Acts 6:1-3 describes how a deaconry was created because of injustice done by disciples in food distribution between Hebrew and Greek widows. Realizing that the church has a technology abuse problem should not be shocking or too hard to fix.

We can take care of our technology misuse ourselves or we will find out that restraint comes in various ways and from various sources. In some areas, power companies vary their rates for electricity which limits consumption during peak demand and controls how we use electric stoves, washers, and dryers. Recently, after years of publicity on the problem of child abuse in church, and after expensive lawsuits, churches had to be forced by their insurance companies to comply with safe child practices for Sunday School classes. If we wait, change will be forced on us to our shame. It will not glorify the name of Christ.

Laws restrict radio and television. The 21st Century Communications and Video Accessibility Act of 2010 requires video descriptive programming on TV. For years, this has been available on tours, in museums, art exhibitions and sporting events. Descriptive video and audio captioning can be technically simple or sophisticated, short or long. It can be a spoken part of an introduction or be inserted during a video clip we download or make with our digital cameras or cell phones for church. Written and narrated descriptions of key spoken and visual elements of a video benefit more than the people who have a vision, reading, or hearing impairment. Such universal design aids everyone's understanding and memory.

History and current experience teach us that concrete obstructions created by new technologies need concrete laws, not merely changed attitudes or a new theology. Barriers must be demolished and new behaviors required. Despite years of disability civil rights laws in America, Britain, and Canada, some churches are still inaccessible. It is worth noting, the younger generation cannot afford to buy new technology or to build and maintain church buildings. Their parents and grandparents pay for this. Misuse of technology is not a generational clash, it is a crime and a sin committed by leaders harming people of all ages.

None of us will hear, "well done good and faithful servant" because we learned to use the latest computer program, unless we first consider the least among us in the way we use it. The church has influence for good and for bad. Our failure to be a light of the world in this area contributes significantly to the ongoing resistance from technology manufacturers, many of whom refuse to use existing, inexpensive ways to make their machines and programs accessible despite the laws requiring them to build accessibility into their products and programs.

Legal restraint is not yet a viable solution to disability discrimination committed by churches because enforcing one case at a time in court with $5,000 or $55,000 fines is too difficult. More easily enforceable approaches should be added with smaller $50 and $500 fines against the misuse of ICT that could be easily reported, proven, and adjudicated. That level of policing occurs now for cell phone use while driving and exists with building safety codes, even for churches. For example, a home-mission team stacked supplies in front of an exit for easy loading in the morning. The city fire marshal warned the pastor and the barricade was removed that night. Disability discrimination could be diminished before it occurs if we had more easily enforceable laws to target schools that teach inaccessible Library Science, or Computer Science, or any inaccessible ICT. We need laws to target and restrict computer program writers, technology manufacturers and Internet sites that supply inaccessible material to churches.

Easily enforceable additions to existing laws against individual acts of discrimination could change entire industries following the successful examples of the National Highway Safety Act, and the Communications and Video Accessibility Act, the Federal Communication Commission's proposed Open Internet Order, or the Net Neutrality Rule which seeks to protect access to the Internet for people who have impairments. Practical pressures and restrictions, though not perfect, are possible and do change behaviors. These exist in many areas and more are emerging, but it is a slow process. Some Bible inspired leadership from churches would help.

Twenty-six states in America require disability awareness training in public schools and colleges now. More will follow. The content of the training is questionable. Churches may not be mentioned at all and the point of view may be that of the institutions rather than being the perspective of individuals facing disability. Disability awareness training, such as using a wheelchair, or wearing mittens, ear plugs, or scratched goggles smeared with petroleum jelly, or other such experiences are counterproductive. They often create pity or fear instead of awareness.

A disability perspective is earned over time. Learning to live with a long-term or a lifelong functional impairment and to avoid or overcome disability discrimination is not the work of a day, a week, or a month. That is why the Bible has so much to say of eternal value about justice and disability. Short of having an impairment and facing disability, the way to be more aware of a disability perspective is to talk with people who live with an impairment and want to end disability discrimination.

The question of when restriction or prohibition is needed, or is possible, affects all technologies. We already make such decisions and act upon them. The idea is not new. We still have many moral conflicts over technology to resolve. Who will believe or value our input on other technology issues if we will not even provide a large print bulletin or make computer projection large?

Churches, that are living examples, could more effectively share a Biblical perspective on such medical technology issues as, what we develop, and who we allow to live, or keep alive and for how long, and with what technology. Controversy also exists in areas such as, conservation and pollution, the disposal of toxic old computers, poisonous pest control, and air, water and land use and the impact of technology on our inner- and inter-personal relations. These conflicts are over the use of technology-- technology we love, perhaps too much to criticize. The Bible and the church have a point of view worth inputting into discussions concerning the use of technology.

Our source of moral truth, the Word of God, is too often left out of this conversation. Temple with Ball (2012) point out that, we, those who are able and those disabled, are made in God's image and Christ is restoring us into his image (Genesis 1:26-31 and 1 John. 3:2). Inquiry into ICT's affect on people or society that considers only the able-bodied distorts that image. For the past 30 years, discussion of ICT in Christian publications and churches was largely about buying it and how to use it, or contrasting styles of music and worship. As a result, today we value technology in the same pragmatic way secular society values it; for efficiency, utility, and fun and we may feel we cannot do without it or change. By ignoring the harm we did and still do and the laws which forbid using ICT in ways that cause disability, we miss the opportunity for repentance and growth. The church could instead guide itself with the godly resources available, the Bible, and deal with its own culpability and responsibility. Then, it will find itself fulfilling its reasonable service as a many membered body (Rom. 12:1-5). The next chapter explores an historic and recurring obstacle which those who love the Church must overcome in order to fulfill that reasonable service.

"Show me your faith apart from your works, and I, by my works will show you my faith. You believe that God is one; you do well. Even the demons believe-and shudder" (James 2:18-19).

* * * * * * *

We should object when people are blocked or forbidden to worship God.. We need to learn to value our own religious freedom and what it means to be citizens of a free country. There is a huge difference between the freedom to create and enjoy culture and living responsibly with the culture we create.

* * * * * * * *

". . . let us love, not in word or speech, but in truth and action" (1 John 3:18).

Technology is not the only area pastors find overwhelming. One Sunday after I spoke in a church, the pastor came to the pulpit and asked for volunteers to clean the church each week which only his wife had done for years.

* * * * * * * *

The Code of Ethics for Congregations and Leadership Teams of the National Association of Evangelicals (NAE, 2015), affirms that we will minimize barriers that would discourage persons with disabilities from full participation.

6. The Institutional Bias

The Bible and secular governments support inclusion and oppose excluding people who have impairments. Why then are many churches still inaccessible? Some churches meet in older buildings that are difficult to modify and some have limited resources, but in most churches that use ICT it can be used accessibly. Something we struggle with is the lure of the flesh and the allure of technology (1 Jn. 2:16). John Calvin (1846) rejected extreme asceticism, but he warned against our becoming so delighted with marble, gold, and pictures that we become marble-hearted.

Self-centeredness, ignorance, and denial all play a role in ableism, that is, the discrimination against people who have a severe impairment. We think everyone is just like we are. We may not know people with hidden impairments, or not know how they function or are blocked from functioning by things we do, and we hate to talk or think about impairments. We fear them and deny they are normal human variations and we reject them as part of who we are in ways that go beyond neglecting to use person-first language.

We may imagine that it is our right to use ICT any way we want simply because we are familiar with it and like it. Or, we may fear that people would leave or not come to church if instead of encouraging messages of our unique selection and gaining what we want we discussed loss, or disability, or presented the compelling weight and beauty of Christ's righteousness and sacrificial servanthood. Or, we may be trying to attract a new, younger generation, but not realize that people with impairments exist in that younger generation too. We also may not know restrictions on technology are common. In order to change how a church sets up and uses its information and communication technology we need to understand how the Institutional Bias operates against those changes.

In a Biblical and Spiritual sense the Church is a living organism; the body of Christ (1 Cor. 12:27) and the bride of Christ (Rev. 21:9), not yet perfect, but making herself ready (Rev. 19:7). The Church is God's family (Eph. 3:15), and His temple (1 Cor. 3:16). Believers are becoming a Spiritual house and a holy priesthood (1 Peter 2:5).

In a natural sense, in terms of its physical plant, the roof that must be repaired and the water, electric and trash bills that need to be paid, and the lawn to mow, church is an organization, an institution. The Institutional Bias obscures individual responsibility and accountability under a guise of "we did it," or "we had to do it," or "we like it," or "they like it," or "they told me to do it," rather than, "I like it" and "I did it." It is often unclear who is really in charge. This spoken or unspoken but fiercely defended power to claim institutional authority to define a situation more than individual attitudes or ableism are part of the less obvious but more significant reasons churches are slow to change to be inclusive.

Many churches celebrate disability awareness month or week and call for inclusion. Christian media publish a yearly story about people with impairments and/or their caregivers, to try to evoke greater inclusion, but the only messages presented are "change the people." Rarely if ever is it "change the institution." Academic disability studies are acceptable as are stories of and pleas for more good deeds or even a new theory or theology–as if the old one were insufficient. A focus on changing the attitudes and relationships of people is permitted, but mentioning changes fo the institution is not permitted or published.

There is no doubt that we as individuals need to be transformed, but our need for heart change should not overshadow attention to institutional behaviors. Two of the barriers the ICF-2 highlights refer to institutional bias; inflexible organizational policies, procedures, and practices and inaccessible information formats. The pastor of a large church exclaimed, "Take that complaint through the proper channels!" which are able-bodied elders using inaccessible tools who see, hear, and read well.

The focus only on individual acts avoids examining pervasive institutional acts or the lack of action. Critics of culture describe the over arching ways tools of mass media and institutions of mass communication shape, control, and dominate Christian culture. How disability discrimination works or the message "demolish institutional barriers" is unspoken and censored.

The censored message is that, "churches are harming people by their misuse of technology." Love does not harm its neighbor (Rom. 13:10).

The only article asking churches to end disability discrimination in church was published in one denomination's magazine 20 years after it voted to end disability discrimination in their churches. That long overdue article reported that most of the 800 churches surveyed were accessible to people who had a mobility impairment (probably due to local building laws), but only 50% provided help for people with a hearing impairment and only 40% provided some access help for people with a vision impairment (Frank and Stephenson, 2013).

The Institutional Bias and technology has many facets and a very long history. It is as old as the telescope developed by Galileo Galilei in 1609 which helped verify that we are not the center of the universe—a revelation that ultimately glorifies God. Likewise the printing press invented by Johannes Gutenberg in the 15th century allowed more people to access the Bible. Our technology today—if used inclusively, could allow access to the Bible for more people. Instead disability discrimination in church is maintained by self-centeredness, a lack of self-control, and a Christian outreach limited by our ignorance of Biblical, technological, and civic standards as well as by our ignorance, fear, and denial of the range of human variation. This operates by institutional behaviors not by individual attitudes or behaviors.

The loud silence surrounding technological barriers in church is as old as the Reformation which erupted due to the stance that the church is right and need not change no matter how much it strays away from God's will or harms people. Institutions habitually resist change, protect their unjust barriers, and reject their own culpability. Missing are calls to change the institutions that by misuse of technology in the name of efficiency or reaching the young disable and exclude people who have impairments. After more than 30 years of secular laws prohibiting disability discrimination change is not urgent or mentioned for churches.

Institutions with "proper channels" intimidate individuals. Insult is added to injury by the use of labels to attack someone who protests unjust discrimination. This kind of tactic, deflecting the message by blaming the messenger, is not new. Ahab, the idolatrous king, called God's prophet Elijah, "the troubler of Israel" (1 Kings. 18:17). Jesus Christ was called "Beelzebub" (Matt.

10:25). Some labels used today are "ungrateful," "uppity," "bossy," and "just one compared to the many," The label "heretic" is still used when mere silence is insufficient to avoid or close the case.

People with hidden impairments may simply be ignored, or they may be labeled lazy, stupid, inattentive, or uncaring. Those who object to even obvious barriers in church may get attacked with the labels used against all protesters, such as, "complainer," or "rebel," and "trouble maker." Advocates who mention disability discrimination get labeled, "confrontational," "divisive," "impolite," "too pushy," "not diplomatic enough," or "not one of us" to deflect their message Messengers are warned of backlash if things change too fast, or they are encouraged to exhibit stoic endurance in the face of injustice, or they are warned against bitterness and offered as its cure "love," and "forgiveness," but without distinguishing between individuals and institutions, or between sinners and sin. David alone cannot battle all the name-calling bullies within our Institutional Goliaths.

Ryan (1971) calls such name calling "blaming the victim," when the wealthy blame the poor for their poverty. Kagle and Cowger (1984) and Rubin and Roessler (2001) use the term "blaming the client," when counselors and social workers blame clients for not improving despite their facing severe and often intractable environments. Barnes (1996) notes how the sources of funding influence scientists and college professors who thus can only blame, or mention problems with "the people" and not with the institutions that pay for their research and teaching. Sanchez and Fried (1997) assert that the right to expose the larger institutional context is a contest of power. In order to protect the status quo of the institution those aligned with the dominant culture inhibit or block the seeking and sharing of information about censored aspects of the context.

The result of this name-calling and censorship, in the case of disability discrimination, is that any mention of its meaning, who it affects, that it is illegal, or even mention of precise remedies, is absent or minimal in sermons, Christian or secular public media, in academic journals, textbooks, or training programs for Pastoral Ministry, Worship, Missiology, Theology, Ethics or in Bible courses. Disability Studies, and Disability Theology label the crime of disability discrimination a "civil rights" or "minority group" model, making it appropriate for study, but relegates essential advocacy to the status of an "optional activity." The result is that disability remains defined as an individual's deficit rather than the result of institutions misusing technology. Then the law leaves redress up to an individual with a severe impairment.

Churches promote charity, but may fail to first encourage justice in God's house (Micah 6:8; 1 Peter 4:17). They cannot sing with Philip Bliss, "Dare to be a Daniel, dare to stand alone, dare to have a purpose firm and dare to make it known." When discussion of ICT barriers in church is absent from pulpits, schools, or Christian media it is absent from our thoughts and prayers.

Discrimination on the basis of disability in our churches is an institutional problem, not a separate cause for a special ministry. A survey of 300 churches found that for 92% of the respondents, the pastor led the church into using computer projectors (Koster, 2005). This is a small sample out of 350,000 public churches and thousands of house churches in America, but Roger (2003) confirms that opinion leaders spread innovation.

One respondent in Koster's survey mentioned being prepared to give large print paper handouts to the members who found they could not see the computer projection well, but during one hymn-sing, the song sheets provided were normal sized print in all capital letters, not large print text (which is 18 point font up to 72 point font and not all caps). If a church's leadership has not had accessibility training or has only limited personal experience with impairments they will be unable to guide a congregation into using ICT accessibly. A pastor, a worship leader or coordinator, a church secretary, or an Information and Communication Technology worker can usually read small text on paper, on a computer, on the Internet, or projected. Church leaders may decide the church does not need or cannot afford 12 point Arial font as its "standard" size so instead use a thin 9 point Times Roman font or less and "see" no need for 10% in large print paper handouts and "see" no need for clearer projected or LCD text because nobody mentions it and it does not impact them. The institution's leaders all "see."

Over the past three decades we have used computers and sound systems incorrectly and we must use them correctly, inclusively. This should be taught in churches, Christian schools, colleges, and seminaries and in two year colleges where ICT is usually taught. Its Biblical roots must be mentioned as part of worship, in church growth material, and within a wide range of materials produced by Christian media, publishers, and organizations. Disability is not a separate topic. Disability and powerlessness--the scandal of the cross–is central to the Gospel and the Church.

The misuse of technology sends a message contrary to the Gospel. Unlike the Holy Spirit, our electronic light and sound effects are controllable and we try to control people with them. Neither individuals nor institutions like being told they are wrong, yet Jesus rebukes those he loves and tells them to repent (Heb. 12:6; Rev. 3:19; Ps. 32:9). Jesus tells us to break the silence and rebuke the offender (Lev. 19:17; Luke 17:3). We miss the mark when we teach forgiveness without repentance; "If you were blind you would have no guilt, but now that you say, "We see," your guilt remains" (Jn.9:41). We must not ignore the offence of the cross or boast in human power (1 Cor. 1:23-28; Gal. 5:11; Jer. 9:23-24).

The next chapter briefly describes another law and a "fear-of-making-waves" barrier we must overcome.

To paraphrase Martin Niemöller:
First they excluded those with low vision from church. I did not speak out because I could see well. Then they excluded those who could not read well. Again I did not speak out since I could read well. Then they excluded those who could not hear well and again I did not speak out because I can hear well. Then they excluded me and no one was left to speak out for me.

"If you were blind you would have no guilt, but now that you say, "We see," your guilt remains" (Jn.9:41).

The "Institutional Bias" enforces the status quo, it says, "you must go with the flow." The "Majority Bias" believes "we're normal and you're not." It believes inclusion is something extra and disability is rare even though one fifth of the population has an impairment and we are all only a second away from becoming impaired. People with impairments come in all ages and in every race and every culture. Ignorance about impairments may seem common, but it is not normal and ignorance about accommodations is often illegal.

I sent this book to several seminaries. One dean claimed it did not fit anywhere in their curriculum. Yet they taught the Bible, worship and homiletics (the art of preaching).

The bad manners that result from the "Majority Bias" gave rise to several **Disability Etiquette** lists of good manners. Some are found Online.

* Don't ask about a person's impairment, whether mild or severe, without permission.

* Don't endlessly discuss an accommodation as if it were some new and strange invention.

* Refrain from unnecessarily sharing your likes or dislikes about a person's accommodations.

* Include the topics of impairments and disability in conversations and Bible study as not unusual.

* Include disability as part of worship, missions, and church life not as a separate issue.

* Set up activities in inclusive ways, such as evangelizing with small and large print material.

* Don't be condescending when providing help it is our basic responsibility, not charity or pity.

* Don't expect gratitude or a celebration of the wonderful (but really normal) accommodations provided in church.

* One large print or Braille Bible or hymnal does not make a church accessible.

7. Advocacy Beyond Groups

Another relevant law combating disability discrimination is, like the ADA, an important part of the context of churches in America. In 1972 a Tv news series reported on abuse and atrocities at Willowbrook and other overcrowded State schools for children with intellectual disabilities. Then, in 1975 the Individuals with Disabilities Education Act (IDEA) made public education, special education, and related services available to infants and toddlers with Intellectual and Developmental Disabilities (IDD) from birth through age 2 and youth ages 3 through 21 and made services available to their families.

A few years after that I took my guitar to a downtown park to sing Scripture choruses and Hymns. Listeners gathered including some who had been released from a nearby closed State Developmental Center. The Lord opened doors for me to sing at local churches and volunteer with their Friendship clubs. I worked in Special Ed and as a supervisor of a Group Home and of sheltered workshops. After earning a PhD in Rehabilitation Counseling I worked as a research scientist investigating the impact of the ADA.

A simple distinction between the ADA and the IDEA is that the ADA covers 61 million adults with severe impairments and the IDEA covers 7.3 million children with learning disabilities, attention deficit hyperactivity disorder (ADHD), intellectual and developmental disorders and others. A conflict between the laws exist. For the ADA, a person may require accommodation or barrier removal, but must be able to participate on his or her own. The ADA excluded individual assistants as an accommodation, but the IDEA often required them.

By far the strongest advocates in church are parents who seek volunteers to help their "special needs" children participate in regular services, Sunday School or Friendship clubs. There is a dislike of the ADA by some of those advocates because it leaves them out. For some, access per the IDEA means to be, with helpers, included, and access per the ADA means to be, "fixed," to function independently. I worked with and studied both laws and believe they compliment each other. They are both needed and relevant to churches.

Most groups, Christian or not, take care of their own if there is a need (Lk. 6:32; Gal. 6:9-11).

For a church, being accessible to the general public all the time is both needed and achievable. Reasonable access is not a mere preference. Accessible, inclusive use of technology in church is essential. Without it, participation in the Word, worship, fellowship, and Christian service is impossible. Again, paraphrasing the Apostle Paul, "if technology makes a message unintelligible to some people, how will they know what is being said (1 Cor. 14:9)?"

People with impairments, like everyone else, have a variety of interests and needs and desire various styles and teaching. They want to try out churches and choose where they will attend. That choice has been limited to where there are few or no Bible believing churches accessible to people who do not see, hear, or read well.

A small, large, or mega-church can be accessible. There is no style of church, not even a military chapel, a house church or small group, that must exclude people with impairments with its technology. If a church does not want to include people with impairments it may take the work of many advocates to open its doors.

Concert halls, sports arenas, theaters, zoos, banks, or medical facilities, schools, restaurants, stores, and trains, planes, boats, and buses must be accessible and their staff trained to provide reasonable access. Why not our churches? Our message is the most important of all and the process to be accessible is not too difficult, expensive, or time consuming.

The purpose of our church technology and music ministry, as with any other ministry is, "to equip the saints for the work of ministry, for building up the body of Christ until all of us come to the unity of the faith and of the knowledge of the Son of God, to maturity, to the measure of the full stature of Christ." (Eph. 4:12-13). This means imparting Biblical, doctrinal truths through sermons, teaching sessions, songs, or individual and group activities. This only works if what we communicate is accessible. It must be heard, seen, or read to be understood, learned, hopefully practiced, even memorized, and by grace, lived out.

Corporate worship is an effectual part of equipping the Saints, any or many of whom may have or acquire a hidden, slight to severe, long-term impairment. If the goal is to attract seekers who are not yet Saints, that group includes people who may have or acquire impairments too. If the goal is to grow like other churches, it is best to emulate a healthy, inclusive church.

Impairment Group: The Apostle Paul personally knew severe, long-term impairment (2 Cor. 12:9). He called for unity in the body because not all parts of the body function the same way, but all are needed (I Cor. 12:14-27). A small portion of those who cannot see, hear, or read well may not read or sing regardless of the clarity or size of print, or may not be able to see, hear, or read at all. Some may simply not want to sing in church.

Whatever the group or their goals we can be accessible. Whether in a more traditional church format with its emphasis on the words of worship, or in a contemporary style gathering with its emphasis on experiencing the physical worship atmosphere, there are people who do not see, hear, or read well who could participate and contribute. They need not all be restricted, segregated recipients of a special outreach ministry. With better use of ICT more people can partake of worship with the church. God is not the author of confusion, but of peace (1 Cor. 14:33).

People with different types of impairments face barriers that must be demolished. At some time everyone may need to be encouraged and comforted and be included, but our goals may differ. Moving to the beat of loud rhythmic music may allow participation for some, but harm or drive away others. Colorful computer projection with small print may attract some, but be a barrier for those with low vision or limited reading ability.

Despite those differences, more people, perhaps 95% instead of 66% will benefit from general changes that improve accessibility in church. There are measures that are easy to learn and to apply whereby we can include more of those whom we may now be excluding. We can extend God's grace though we may not reach everyone or even know who we do reach.

Environmental conditions increase or diminish the severity of an impairment (Heb. 12:13). This is a fundamental moral value of the Rehabilitation profession (Wright, 1981). It applies to both physical and social conditions and became known as the Social Model of Disability. Laws against disability discrimination require that we, whenever reasonably possible (1) remove physical barriers and (2) apply universal design, which means we make things and activities so people who have impairments can use them and (3) provide reasonable accommodations (called adjustments in Britain). These steps allow for the maximum use of a person's abilities.

Unity of Advocates: Since those three steps may not always equally help everyone with any impairment in all situations, the Social Model and the laws created two groups. One group is helped by changes to the environment such as the ADA might require, while the other group may need one-on-one help from volunteers or paid staff such as the IDEA might require. That group may fear making waves, preferring a more nonoffensive, social interaction over advocating for accessibility changes. The terms "inclusion," or "person with a disability" for that group often refers to people with intellectual and developmental disabilities (IDD). Their caregivers may prefer seeking social acceptance and better relationships and changing attitudes which they do need over the risk of alienating people by advocating for inclusive print, projection, or sound which they may not need.

The debilitating nature of the fear-of-making-waves barrier is highlighted in John 9:19-22. Synagogue leaders asked the parents of the man born blind to verify that the one Jesus helped was their son and to tell them how he was healed. His parents were afraid they would lose their position in the community so they owned their son, but did not own Jesus. Their son confessed Jesus was working in his life and for confessing that fact he was cast out. Fear of loss of community or loss of needed assistance due to one's own or a family member's disability still divides disability advocates.

For both groups advocates are involved. Often it is the parents, not the person who has a disability that have a strong, well-known voice, but they may only know about the needs of those they care for. People with various types of impairments benefit from the combined efforts of many advocates even when they have different goals, especially if the loudest or the only voices heard are the caregivers or parents of people with only certain types of impairments. A fear barrier can keep advocates silent, separate, and even on opposing sides of issues.

The caregivers of people who have intellectual developmental impairments or who have multiple profound disabilities who are not helped by accessible ICT may not know about technology barriers. They or their son or daughter may not want to read along, or sing along. Nevertheless, advocates for people with disabilities who are not affected by the way a church uses its technology could support the efforts of those who do care about accessible use of technology.

We are united by a Gospel that changes insensitive hearts and by the fact that incorrectly used technology may create barriers to people who have a vision, reading, hearing, intellectual, emotional, developmental, or physical impairment. Some people with any type of disability in a church of any style are helped by accessible use of information and communication technology. We do not have to choose one group or another. We can be advocates beyond impairment group or type as followers of the Messiah who helped people see, hear, and understand the Word of God, who, on the cross, was disabled and died for all.

Accepting people with impairments just as they are without demanding they be "fixed" by medical technique or prayer intervention, requires that whenever we can, we make our physical and social environments accessible. This includes how we use technology. It will not help everyone. Some would not want to come to church no matter what we did and a small percentage could not see, hear, or read no matter how we use our technology, and some spaces and technology cannot reasonably be changed. This occurred in the first century too. At that time in the Holy Land, before glasses were invented, many people who could see, hear, or read some, but not well were called blind, deaf, uneducated or illiterate.

Removing barriers, using universal design and reasonable accommodations and adjustments all without undue burdens or altering our essential purposes can help some people with any type of impairment and in any style of church. It is not a question of us or them, here or there or helping or harming this or that group or trying to help only our own or only the neediest. As this often used play on Frost's poem says, "Two roads diverged in a wood and I took them both."

Consider how King David dealt with the exhausted third of his army after God gave him victory following the destruction of Ziklag (1 Sam. 30:9-25). The army rescued all the captives and recovered far more spoil than anyone had lost, but among the two thirds who were strong were those whom the Bible calls corrupt and worthless fellows. They did not want to share the blessings God allowed them to acquire with the exhausted one third who were not able to keep up. David made it a law in Israel that all in his army, weak and strong, would share alike. We too can reach for that goal in our churches.

The next chapter lists several more common and often encountered barriers we need to bridge.

Accessibility does say, "Welcome Visitors," but hospitality is for guests or strangers. Many people with impairments belong to our church or are our friends or family members. They are not strangers or visitors and do not want special hospitality. They expect appropriate respect. Most Millennials and GenZers grew up with accessibility and special education in their schools. They may well wonder why the church excludes so many people with impairments.

8. Am I My Brother's Keeper?

Some churches welcome people from assistive living centers or group homes for people labeled IDD or people learning English or people who do not see, hear, or read well, however there is opposition. A bully in the schoolyard may start a fight by calling a student wearing glasses "four eyes." There is also a bully in the churchyard. My impairment is often unobservable so I can pass as "normal" and enter most groups. Thus, I hear the conversations of those who use the word "ugly" to describe wheelchairs, large print, clear projection, a hearing loop, or signs that indicate accessibility.

Cain refused to admit he had caused harm or that he was responsible for the harm he caused (Gen. 4:9). His testimony illustrates two facets of denial that keep pulpits and classrooms silent on this issue despite the rich Biblical record concerning people with impairments and despite the laws prohibiting disability discrimination which are the context and environment of churches. In addition to ignoring the communication barriers we create with our technology or the harm we cause with it, the following are some often repeated ways and reasons given for not being an accessible church.

1) As mentioned in chapter 4, obstructions to being an accessible church begin and continue by the failure to teach Christians the Biblical understanding of human variation, or laws against disability discrimination or a disability perspective.

2) Further, obstruction comes by folding the ADA into all outreach to people with any special need instead of setting up church in the first place with no institutional barriers before anyone shows up. This is denying the fact that someone is misusing the ICT now. One way this barrier appears is to agree to the easy and obvious, but delay using technology inclusively until more volunteers or staff have training and the church has a plan to deal with all accessibility needs even the most difficult or expensive. Avoiding the easy by aiming for the hardest allows leaders to sound as if they are concerned for the least of the lost. A church may get a sign language interpreter or make Braille translation, which are good things to do, but they reach the fewest number of people, usually those who do not hear or see at all. Then in a short while, when no new people show up, the whole accessibility plan gets abandoned.

A wiser approach is found in Proverbs 3:28, that is, do what you can do when you can do it. Do not wait. The methods for using technology accessibly have been available for decades and are easy to learn and to apply. The Biblical perspective in chapter 2 and 3 can be shared now to change hearts and actions.

3) Another barrier is foisting conflict between the ADA and the IDEA as if caring for individuals is in opposition to creating safe, inclusive, non-disabling environments. Both are needed. This conflict may arise from the divide between State Rehabilitation Services and public school services under IDEA. Rehabilitation services help people with severe impairments access full time work or college, but not with help from assistants or volunteers and it does not help people work in below minimum wage jobs at sheltered workshops. The church can be accessible to both groups and advocates for either group can help support the others.

4) A huge barrier is the fallacy that being willing to provide access >in case it is needed< is the same as being accessible which is always needed.

5) Another barrier is the claim that a majority of the church does not need access when the minority is already mostly excluded and has no voice except for unspoken scripture. One church declined to be accessible because it affected only two percent in their church. That was 8 out of 400. Eight people were in Noah's ark.

6) A critical barrier, mentioned in chapter 4, is the misconception that required access is the same as a choice of good manners or hospitality. It is not.

7) People without technical knowledge cannot refute biased barriers on the false claims that lack of accessibility is due to the technology itself.

8) Other barriers requiring insider or technical knowledge are misleading claims that lack of access is due to a lack of trained volunteers.

9) A common barrier is challenging the style or tone of the message or messenger or changing the topic to anything else, but not engaging with or discussing the content of the ADA message. Church leaders are trained in communication and may easily overwhelm persons with impairments.

10) An unspoken barrier is to display technology instead of being a display of Christ followers. This claims technology, like coffee and donuts, brings people to church. This asserts the power of the Gospel is unattractive without bells and whistles.

(11) The technology myth says a technological fix or cure will come if we wait. Some new, do-it-yourself technique will emerge so the majority will not need to change its ways for the minority.

Unfortunately, computers and the Internet today are becoming less accessible for people with impairments (Frank, 1999; 2006; Blanck, 2014). More barriers to Web equality are being faced by people with disabilities even when solutions exist. "Let them take care of themselves" sounds as easy as bringing your own earplugs to church, but it is not easy and it does not encourage our meeting together (Heb. 10:25). The message sent by a "make-them-do-what-no-one-else-has-to-do" approach is, "stay home!"

(12) "Send it to the committee" or "do more research" are well-known ways used to kill or delay action on something that can begin immediately. That ruse claims there are no clear problems and solutions and instead calls everything an opinion.

Scriptures and research on disability discrimination are plentiful, but this is rarely if ever shared in church. A committee may need to start by learning the difference between a special "disability" ministry and applying self-control and obeying the law when using technology for "everyone."

(13) One way research is used to block access in church is to survey the congregation to see if enough people want the changes. Besides ignoring the Word, the law, the larger community, city and country and the future, church members might not say they need help in order to avoid conflict or they may not know how easy it is to be accessible. Another misuse of research is to aim only for the young because one survey found about half of Christians said they were converted at a young age. This overlooks the record in the Bible and in church history and excludes the other half saved at an older age and those adults still needing to be reached and nourished.

Research is also misused by asking church members with a hearing loss to vote on a hearing loop, but they have no experience with that type of hearing assistance system and are given an exaggerated price quote. Not surprisingly, it is voted down. We may falsely think an uninformed, misinterpreted, or slanted survey is an attempt at research, or democracy, or we may believe the majority rules, but in a modern democracy where the majority rules, it does not get to invent the standards and the minority is protected.

After promising to install a $2,000 hearing loop when the church bought a new sound board, a Pastor apologized for forgetting to buy it. Nobody believed him, but they had a new sound board. In another church, the price given for a hearing loop included, without mentioning it, the price of a new carpet for the sanctuary presumed necessary to hide the "ugly" hearing loop wire. Enough people who used hearing aids believed it to vote it down.

(14) A related barrier is to consult retailers on ways to improve sound or computer projection. They examine the facility and suggest buying a better sound board for $10,000, or a stronger computer projector for $6,000, or an LCD screen. The false conclusion is given that accessibility is too expensive. Retailers may say the problems can be reduced with curtains or sound absorption material, but they are unlikely to suggest lowering

the pitch of music, or increasing font sizes and spacing, or projecting fewer words on each slide, or improving the letter/background color contrast of an LCD or projected text. They may not even be aware of those steps. The lure of selling and buying new technology makes a new computer and a plug-it-in-and-push-one-button solution attractive. Unfortunately, this may only lead to a brighter and louder church, but not greater clarity.

(15) A sad obstacle to an accessible church is forgetting that God shows no partiality (Acts 10:34-35; Ro. 2:11). The idea that nobody here is disabled, that this is only for those other people somewhere else, reveals a self-centered Christian church. The prevalence and type of impairments vary between regions, races, income levels, gender, and age groups, but limited vision, hearing, or reading ability affect various types of people everywhere. The Bible says, "There is neither Jew nor Greek, there is neither slave nor free, there is no male and female, for you are all one in Christ Jesus" (Gal. 3:28).

(16) Sadder still is raising the fear that valuable relationships could not or would not develop even

with better access in church. In response to that objection, I confess that I too have participated in conversations at church with no personal or spiritual content. It is easier to talk about topics commonly heard in secular environments than to engage others in deeper spiritual or personal issues.

I can recall church suppers where each table was filled with only one family and there was no mixing at all. People with impairments were not the only ones left out or segregated. I can think of small groups that discuss their latest family or work news, but they no longer find Bible study appealing. I also know first hand the chill that can and does come into relationships when disability discrimination and equality of access in church are mentioned. The fear-of-making-waves is real.

Few of us are popular, extroverted leaders, whether able-bodied or not. We mostly just try to fit into the social structure around us rather than to create or change it. But Jesus said, "By this everyone will know that you are my disciples, if you have love for one another" (Jn. 13:35), and "it is more blessed to give than to receive" (Acts 20:35). Note too, "A man who has friends must himself be friendly" (Prov.18:24 NKJ). Access is a precondition for participation in church. It is not the end point. All of the Gospel must take root and grow. Jesus said, "if I be lifted up, I will draw all peoples to myself." (Jn. 12:32, NKJ). As we draw closer to Jesus we will draw closer to each other.

17) A piano player who is blind, but knew the entire hymn book by heart told me someone from the choir would pick her up for practice, but they would all often forget to drive her home, leaving her stranded at church. They could not imagine someone not driving a car in their rural area.

18) Parents in a church could not control their children who liked to play with the automatic door openers and the elevator so, mindful of the cost of repairs, the church elders shut them off and posted out-of-order signs. One lady who used a walker contacted the city government and got the door "fixed." Someone else complained to the secular authorities about the false sign on the elevator and got that restarted. Walking around the first floor to find the person with the key was not easy for a person with a mobility impairment. The elders erroneously claimed they knew everybody who legitimately needed the elevator.

19) People complained about confusing colors, pictures, and movement on the projected song slides. The worship coordinator made clear slides with good color contrast, but only for the evening service. This created two churches, the morning for younger people and the evening for elders.

20) One church decided to do nothing until after they remodeled the sanctuary which was unrelated to the changes needed and was years away..

21) Even opponents of abortion and euthanasia fail to see the contradiction between their pro-life beliefs and misusing ICT. Some people with impairments are identified in Utero, before birth, and are aborted. Others, abandoned and alone when older, choose suicide.

22) Even opponents of evolution may embrace social Darwinism and survival of the fittest, acting like and even saying, those who are unfit (as they define "fit") should not be in church.

23) Denial and avoidance disguised as religion are difficult obstacles. At times, as a hospital chaplain, I encountered people who insisted on

success only as they defined it or held some other health and wealth prosperity doctrine. They denied disability could be part of a Christian's life and refused to believe any severe impairment could happen to them or to their loved ones. Some patients and families deflected every question about disability and their path forward. They could not or would not deal with the details of living with an impairment even though the patient could not be released without a safe place to go.

24) Barriers may exist because of underlying philosophies and personal motives, notably, the fear of impairment and its potential for dis-ability For many, a long-term impairment is the thing we fear the most (Job 3:20-25). We project our fear onto others and assume they, like Job, are miserable. The truth is, many of us with a long-term impairment live full lives--except for the disability discrimination we face in churches or elsewhere.

25) Another barrier is the anti-law (antinomian) position which exalts grace or freedom as the right to do whatever one pleases or thinks profitable, rather than as the opportunity to do what is right. Lawlessness ignores the Bible's view and invents

fault with laws that restrict harmful behaviors. Resisting government intervention in churches may at times be justified, but not as an excuse to ignore secular regulations of ICT while at the same time we imitate secular excesses with ICT.

26) Access is thwarted by "dancing around the issue" or by "polite" silence about disability or by denying accessibility standards or medical, legal or educational standards exist, such as, the range of variations, slight to severe, between normal vision and hearing, and total blindness or total deafness or the range from illiteracy to literacy. Confusion is created by asserting that generic standards might exclude individual needs or that they are only subjective opinions or that every and any limitation or only extreme limitations are an impairment being turned into a disability.

27) Another confusion is framing disability as a contest with winners and losers in a competition, as if in a sporting match. This rejects the virtue of fair play and reveals a "might makes right" ethic that leaves us back in Judges 21:25, where everyone did what was right in their own eyes.

28) The belief that computers are, or should be easy to use, but fearing mistakes in the process of learning to use them accessibly, is yet another contradiction we live with.

29) Our perseverance and frugality are tested when no one uses the large print we provide in advance of requests. But this happens because regaining people's trust takes time. The goal remains for us to cultivate the wisdom and spiritual fruit of self-control in an environment where access for people with impairments and knowledge about providing access is required by Biblically based, but poorly enforced, secular laws. The bully in the churchyard wants to hinder worship to God (Isaiah 14:12-15; 2 Cor. 11:14).

30) Another deflection is claiming that disability is only about voluntary, individual relationships or is just, another "cause" or charitable activity. This frames **the disability we cause by misusing ICT** as one more choice or interest group among many. But this is about us-we who weekly commit disability discrimination that limits our mission, our one cause—which is to communicate the Gospel. This is not another cause. It thwarts our one cause. It is illegal and a sin and can be easily prevented.

And the list goes on. In a church which I helped find a large print Bible, three leaders, the worship leader, computer technician and pastor rejected accessible ICT use because they "like" the way they used it. Ignoring the needs of others takes planning and work. People abuse alcohol, tobacco, and opiates and eat too much sugar, salt, and fat because they "like" it and it is easy to do.

Not long ago, the air in many church's coffee rooms was blue from tobacco smoke. We learned that was harmful and changed. We can change our harmful misuse of or addiction to information and communication technology. Restricting abuse is not legalism it is love. It takes planning and work.

Misuse of ICT is systematically carried out as churches are set up every week. People with impairments of any age may already know from experience not to come to church. They only have to look around and listen to know if a place is accessible. They are not likely to tell us about their hidden impairments. We have our work cut out just to get our neighbors to visit again. Our missionary outreach may also be unnecessarily inaccessible because someone "likes" it that way.

* * * * * * *

Determining the exact number of people who need inclusive access to ICT in a church or community asks, "Who is my neighbor?" instead of asking as Jesus taught, "To whom am I being a neighbor (or not)?" Nonetheless, in order to learn more about what it means to be human and as was promised in the introduction to this book, the next chapter offers disability demographics for America, Britain, Canada, and the world.

9. **Who Is My Neighbor?**

Demography is an old, imperfect science because people are involved. We prefer easy-to-read data, but social science researchers know all too well the difficulties and the expenses of collecting and reporting data on people with impairments. We use what are considered the latest and best data sources, but without claiming omniscience and there is room for improvement.

Multiple data sources are presented here in order to acknowledge variations in the reported dimensions of these populations. This may seem confusing, but there is not one easy number to report for each population. The reports from different data sources do not question the large numbers of people of all ages with seeing or hearing impairments, or limited reading ability.

America: The American Foundation for the Blind (AFB, 2012) estimates that 20.6 million Americans cannot see well even with glasses or contacts. The National Institute of Literacy (NIL, 2013) estimates that 40 million Americans of working age find even a basic literacy level (11 year old, or 5th grade level) difficult. The Hearing Loss Association of America (HLAA, 2014), citing

Center for Disease Control data, estimates that 48 million Americans have some degree of hearing loss. These estimates cover people with slight to severe impairments, including a small percentage who cannot see, hear, or read at all. Inclusive use of technology in church could help many, but not all of these people. Not all would want to attend. Some are members or already attend church now even though they cannot fully participate.

Some surveys do not specify the population. Some omit people living in institutions and some refer only to people between the ages of 16 or 21 to 65, not to the entire U.S. population This omits 130 million people. In 2020, the U.S. population was 332.5 million (census.gov), The Centers for Disease Control and Prevention (CDC) (2021). reported 6 percent of adults or 61 million had functional disabilities. Of those, 4.6% had a vision and 5.9% had a hearing disability. An earlier survey (SIPP), found 5% of adults had difficulty seeing, and 1.9% had difficulty hearing (Brault, 2012).

Consider some other sources. The National Institute on Deafness and other Communication Disorders (2005) estimates that 30 million people in the U.S. age 12 or older have hearing loss in both ears (not 6.2 million per SIPP or 48 million per the CDC). The AFB cites the National Health Interview Survey data that 20.6 million people have difficulty seeing even with glasses or contacts (not the 15.4 million per SIPP). It is not always clear why estimates from different sources differ. The intended purpose of the data, the questions asked, or the definition of a disability, or the ages considered, or the year of a survey, or various characteristics of survey respondents or those who create or conduct surveys can each affect results.

Large variations, a low or high estimate, do not mean the information is invalid. We choose data sources for a reason and we include people in church for a reason. If employment issues are the reason then ages 21 to 64 make sense. If church participation and respect for parents and for the elderly are the purposes we ought to use data that include all ages (see Eph. 6:2-3, Lev. 19:32; Ps. 71:9; Is. 46:3-4). The combined incidence of all impairment types reveals that 20% of people under the age of 64 and 40% of people over the age of 65 have a disability (CDC, 2021). Do our churches exclude and dishonor people over the age of 65? In some places they outnumber those under 18.

Gallup polls typically find about 40% of the American population say they attend church weekly. The numbers vary by region. The exact number of churches in America that misuse ICT is unknown. The precise number of people of any age who cannot participate in churches that misuse technology is unknown. Survey and anecdotal data suggest 60% to 75% of the 350,000 churches in America misuse ICT and unnecessarily disable and exclude as many as 80% of those who have impairments and their families. When differences in data exist shall we abandon compassion and, contrary to Abraham say, "Lord, we will not be inclusive because, peradventure there be not 50, but only 45, or 40, or 30, or 20 or 10," (Gen. 18:24-32). What if there was only one more person we could include, instead of millions more?

Britain: The total population of the UK was almost 67.9 million in 2020. Earlier data reported just over one third (34%) of adults living in Britain said they had a long-standing illness or disability (LSI). Almost one fifth, (19%) said they had an LSI that limits activity (National Health Statistics [NHS], 2013, Office of National Statistics [ONS], 2012). There were over two million people in the United Kingdom with sight loss in 2015 (National Health Statistics [NHS], 2015). The number was expected to increase to 2.25 million by 2020 and double to four million by 2050 (Stade, Royal National Institute of Blind People, [RNIB], 2014). Almost 3.5 million people of working age (16 to 65 years) are "deaf or hard of hearing," and ten million people in the UK have "some degree of hearing impairment or deafness" according to Action On Hearing Loss (2014). Almost 16% of adults, or 5.2 million are described as "functionally illiterate," that is they have literacy levels at or below those expected of an 11-year-old in the 5th grade (National Literacy Trust, 2011).

The combined data for adults between the ages of 16 and 65 in the UK indicate slightly more than 17.2 million reports of a vision or hearing loss, or low reading ability. These could overlap with some individuals having more than one sensory impairment and/or limited reading ability. The number could be nearly twice that if all ages were included. The number of immigrants in the UK is large. Illiteracy is more often an educational deficiency rather than a learning disability. Some immigrants are from countries that deny education to females. Again, this is not 23% of the entire population of the UK. It is neither 17 nor 34 million individuals, but rather, 17.2 million, or, if all ages were included, twice that many reasons to use information and communication technology in accessible ways in all 50,000 churches in the UK.

Canada: The population of Canada In 2020 was more than 37.9 million. Earlier data reported 3.8 million adult Canadians age 15 and older, 13.7% of the adult population, reported being limited in their daily activities as a result of disability. This includes 3.2% of the adult population having a hearing loss and 2.7% having a vision loss (Statistics Canada, 2014). In Ontario, with a population of 13,135,000, 1.8 million adults report having a disability.

The Canadian Association of the Deaf (CAD, 2012), with strong disclaimers as to the accuracy or dependability of any data, reports 350,000 profoundly deaf and deafened Canadians and possibly 3,150,000 who are hard of hearing. The Canadian Hearing Society (CHS, 2013), reports nine million adults (almost one-fourth of the population) have some hearing loss, with one million having a hearing related disability. According to CHS, some studies indicate the true number may be three million or more adults with a hearing related disability because those with hearing problems will often under-report their condition.

There are substantial differences between "some hearing loss" in nine million adults, and an "under-reported hearing related disability" in three million adults, and reports of "a hearing related disability" in one million adults, and 350,000 who are "profoundly deaf." The ages included, the ranges and definitions of what is measured, and the accuracy of responses affect survey results, but the numbers are huge. The 2012 Canadian Survey on Disability (CSD) by Statistics Canada found 1.1 million adults (1,137,292) with a hearing loss in Canada with 437,718 in Ontario (Employment and Social Development Canada, 2015).

The CNIB (2007, 2013) reports that half a million Canadian adults have a vision loss. It is not clear if that means (in the same way as the hearing loss data), "some loss," or if it means enough loss to be at a level of a "vision related disability," or if it means total or near-total blindness. This data may also be incomplete due to under-reporting, but these qualifiers are not included with the Online CNIB data reports.

On the other hand, according to the 2012 Canadian Survey on Disability (CSD) by Statistics Canada, there are almost one million (959,590) people with a vision loss in Canada, including more than a third of a million (369,324) in Ontario (Employment and Social Development Canada, 2015) or almost twice the number reported by CNIB in 2007. In addition, for the first time in Canadian history there are more people who are age 55 to 64 than age 15 to 24 (Statistics Canada, 2014). As previously noted, while the incidence of disability increases with increased age, impairments exist in people of all ages.

It is understandable that people would under-report a hearing or a vision impairment. People who do not hear or see well might not respond to a spoken or written survey consistently. Problems exist in researching and reporting demographic data on the incidence and severity of impairments. The sources of data, the Participation and Activity Limitation survey (PALS) and the Canadian Survey on Disability (CSD) are not equivalent surveys and cannot be compared, but the highest and lowest counts are large.

Comparing the data from different countries and within the same country is made even harder because the definition of adult varies in the data, (15 to 65, or 21 to 64) and the cost and methods of surveying and testing, and the meanings of the terms "loss" or "disability" vary. Further, the consequences of being labeled as having limited hearing or vision vary between countries and within countries. There are incentives for under-reporting, such as, the fear of losing privileges, or a lack of benefits such as, helpful medical or rehabilitation services or supports. Still, the data indicate a large population with hearing and vision loss who could not fully participate in any of the 30,000 churches in Canada, not even the 11,000 that consider themselves Evangelical whenever the churches misuse information and communication technology.

There are also huge numbers of people who do not read well in Canada. In 2003, 42%, or 16 million adults, age 16 to 65, had low literacy skills (Canadian Literacy Network [CLN], 2015). In 2012, almost half, 48.5% of Canadians, age 16 to 65, had low literacy skills with scores in the Level 2 category or lower; that is, a 5th grade or 11 year old level or lower. In the later primary grades tracking, scanning, and fluency can be adversely affected by the smaller fonts and the increased concentration of letters, words, and lines on a page. In Ontario, 47%, or 6.2 million adults had low literacy skills (Employment and Social Development Canada, 2015).

Canada is a bilingual nation with a large number of immigrants. The reported decrease in literacy may be largely attributed to increased immigration from countries where the opportunity for an education is limited. Those factors are taken into consideration by CLN which offers reading tests in more than one language.

In 2011, according to Indicators of Well-being in Canada (2012), disability services were needed by 12% of the college students in Ontario. As of 2012, almost 25% of adults in Ontario have a university degree and almost half of the adults in Ontario cannot read well. Whatever the reasons for this literacy and education gap, there is a huge need for accessible projection of Christian materials, and large print materials, and especially, truly large print Bibles.

There is little point arguing over demographic data as though making a cost/benefit estimate. We can instead use the best standards for readability without trying to simplify or dumb down the material. Church communities consist of people of all ages who have a variety of strengths and weaknesses. We cannot assume any group, young or old does not need to or does not want to participate in church. Our information and communication technology today allows us to easily create larger print, better color contrast, better letter, word, and line spacing, and fewer words per page or slide. We can lower the pitch and add a hearing loop or FM system to our sound systems even if only one person needs it. Jesus asked, "Which of you will not leave the 99 to go find the one?" (Matt 18:12).

Worldwide: The World Health Organization (WHO.org 2020) of the United Nations estimates that 15% of the world's population of 7.8 billion has a severe impairment. If those almost 1.2 billion people were in one country, it would be the third most populous nation after China and India. They do not all have a vision or hearing loss. Based on countries which collect such data, 314 million people worldwide have a vision impairment and 360 million have a disabling hearing loss (WHO.org, 2010). In addition, 775 million adults worldwide lack minimum literacy skills (UNESCO.org, 2014).

Language translators work for years creating a Bible and Christian literature in the native tongue of unreached people groups, but the technology that could produce various print sizes is often used to reproduce only one size of print books, or papers, or video or projection for a people with limited or no access to optometrists or reading glasses. Audio material and internet sources help some, but they cannot replace books and paper.

From Roman roads to satellites advances in technology have opened doors for sharing the Gospel and they can do so even more if we use our missions resources inclusively. The key is to use our technology as a bridge to include 95% of the variations in the population, not just 51% or 67%. We can plan to be barrier free in the first place. We need to humbly learn and teach that not everyone is an able-bodied, highly literate student, teacher, minister, or missionary.

The focus needs to be on building our environment and using our tools in accessible ways–not the number of people with impairments –and not only one at a time. A self-evaluation, accessibility audit can reveal what we are doing wrong and guide our corrections. Thirty years ago we were limited by print books and typewriters. We have the technology today to set up church and outreach ministries in easy and inexpensive ways that could reach many more people.

The moral fiber and character of a nation come from the Bible and from the churches. We can lead and be a light in the world rather than waiting until others force us to do what is right. Most of us object to being told we are wrong, but before asking, "What extra thing can we do to help those we inadvertently left out?" we can first ask ourselves, "Are we being a neighbor in the social, physical, and technological environment we create?" We are called to help others, but we can also avoid harming others in the first place by considering how we can use our technology more inclusively all the time.

There are of course limits to what we can do. Impairments range from slight to severe. The generic accessibility standards listed in the next chapter will not help everyone and individual help for some, when it is available will still be needed, but it is not necessary or normal to be cut off from Christian community because of limited vision, hearing, or reading ability. When a church is inaccessible because, "we didn't know," or "nobody here needs that" (yet), it is likely that people with impairments who were ignored did not say anything. They just participated less or left. They may have chosen to leave for a variety of reasons, but if disability discrimination was a reason people left we may have chased them away and we have work to do to bring them back, to reconcile and restore. Hopefully, when people learn that a church is trying to be accessible, the community will help to (1) remove barriers, and (2) use universal design, and (3) provide reasonable accommodations and adjustments.

Bible believing churches can lead the way in being accessible because this is central to the Gospel, it is not something extra we should do. Those who rebuked blind Bartimaeus changed their tune when they heard Jesus call for him (Mark 10:46-52).

Is technology god, or can we–will we, change the way we use it? The previous chapters on the Bible, the laws, the culture, the name-calling, the arguments used against being accessible, and the huge number of people negatively affected by misuse of ICT may shock people unfamiliar with 21st century disability discrimination. Ignorance and resistance to inclusion are precisely why the scriptures, the laws, and this book are needed.

A church presented a missionary's newsletter in its bulletin, not in standard size or small print, but in tiny print. A denomination headquarters distributed brief biographies of their missionaries on glossy paper in tiny print. They fit all the information in, but it was extremely hard to read by people with normal vision, not just by those who did not see or read well. They mistakenly thought they were furthering the cause of Christ.

The next chapter shows what and how we could change if we have ears to hear what the Spirit is saying to the Church.

Beware of Geekspeak: If someone claims the church computer's CPU and RAM, or its hard drive memory are too small or too old to project or print large print slides or paper, point out the weekly super-large print slide, "Offering Time" or the huge computer created paper sign in the entranceway, "Support our Missionaries." Small or large print, or projected or LCD slides can be produced on almost any computer. When a new computer is bought and the choice is again made to use a church's ICT in inaccessible ways, remember, "it's a poor worker who blames his tools for his mistakes."

"If a brother or sister is naked and lacks daily food, and one of you says to them, "Go in peace; keep warm and eat your fill," and yet you do not supply their bodily needs, (*or their access to spiritual food*) what is the good of that?" James 2:15-16 (words in italics added).

* * * * * * * *

"We who are strong ought to put up with the failings of the weak, and not please ourselves. Each of us must please our neighbor for the good purpose of building up the neighbor. For Christ did not please himself" (Romans 15:1-3).

10. A More Excellent Way

For many churches, computers and sound systems are recent additions of only the past two or three decades. We can change how we use them. The best access is not based on opinion or preference. Choices exist, but the most readable formats for computer printing, photocopy, signs, projection, or LCD screens have been researched over many years. So too, the best accessibility for the Internet, Websites, e-mail, and other social media, or smart phones, boards, TV, and video is known. In addition, a variety of hearing assistance devices are available as are various accessibility features and devices for sound systems.

The following guidelines are not meant to suggest a church begin to use technology or upgrade what it owns. A church can apply these accessibility standards to the ICT it already owns, with little cost or effort, at a reasonable pace. The more expensive items mentioned are a hearing assistance system added to a sound system, or a Braille embosser added to a computer. Those can cost as much as a computer system, a computer projector, or LCD screen.

These standards come from various sources such as, the American Foundation for the Blind (afb.org), the American Printing House for the Blind (aph.org), the Royal National Institute of Blind People (rnib.org) of England, the CNIB (cnib.org) of Canada, the World Wide Web Consortium (www.w3.org), and the U.S. Access Board (www.access-board.gov), and the U.S. Justice Department (www.ada.gov). Most manufacturers do not mention how to use their products or programs accessibly, but they might answer phone or e-mail questions on how to do something specific, like increasing font size, or letter and line spacing. Libraries and colleges may answer questions about accessibility or using a program.

These lists are not legal advice or instruction on how to use software, printers, projectors or other technology. An Internet search may help find how to use a particular program or find advocacy groups or standards for accessibility for people with other types of impairments. A 53 page, general introductory overview discussing the inclusion of people with a variety of types of impairments (Davie, 1997, 2016) can be downloaded free at www.aapd.com/wp-content/uploads/2016/03/That-All-May-Worship.pdf.

The following lists of standards will help many people who do not see, hear, or read well, but it is also a good idea to ask individuals what they find useful because people vary and impairments range from slight to severe. Other barriers may exist or emerge, such as lighting, seating, architectural issues, or the introduction of newer technology. Signs, such as bright color strips on the edge of stairs, a stage, on an incline, or over electric wires connecting ICT are usually mentioned in regard to accessible architecture which needs its own book.

This is the third updated version of this book. Changes have been made, but more are coming. That is the nature of technology. It cost some and took some effort to learn to use in the first place and it still does and will in the future. A pastor or church may use the latest Internet or MS Word program. Some older programs translate newer versions, but many people with older programs are not able to read or respond. Each person and church may consider and decide when and what to change or add, but it contradicts and violates the Bible and the laws of our countries to exclude people who do not see, hear, or read well.

1) Printed Material: Computer programs, templates, and printers may automatically compress or reduce font size, or type, or letter, line, and word spacing. Performing accessibility quality control edits before and after printing may help find and reverse this.

Font size and style:

- Large print text is defined as 16 point font by the RNIB, 16 to 20 point font by CNIB, and 18 point font by the AFB. 18 point = 1/4 inch, (18 X 4 = 72 point font = 1 inch). Avoid fonts that are condensed or if 18 point does not = 1/4 inch.
- Average print size is 12 point font, but too often in thin Times New Roman font. Use a wider, sans serif font.
- The RNIB suggests a minimum "clear type" standard of 12 point Arial font as "normal," (not large) text.
- It is difficult to see and read bulletins or other material in a thin font style or in less than 12 point font size.
- Readability is best with only one style and size of a plain sans serif font, such as Arial, or a larger font type such as Verdana, and without adding appearance affects or condensing or compressing letters, words, or lines.

- Expanding letter, word, and line spacing to 1.1 or more may be helpful. Aphont, an expanded font created by the American Printing House for the Blind (aph.org), can be downloaded for people who do not see well.
- Some sources use the term "larger font" for 14 to 16 point font and use the term "enhanced large print" for 24 to 36 point font (Kitchel, 2004).
- If 36 point font is not large enough, audio or Braille is recommended. Larger than 36 may be harder to use, but even one inch, 72 point font can be used, though it may not help everyone.
- Some sources suggest saving from 1% to 30% of printing costs by using smaller, thinner fonts, but this does not save money or souls.

Formatting:
- Reading is easier if each large print page contains meaningful material that makes sense rather than having partial sentences or bits left-over on a page.
- Instead of using a new or larger page when a few words or lines are left, reduce the size of margins, or large headings, or blank space.
- An 8½ by 11 inch sheet of paper filled by 12 point text can be reformatted into 18 point, Arial font, and fit onto an 8½ by 14 inch sheet of paper.

- Large print handouts should be created and folded so as not to look or feel bulky. If they are cumbersome or feel too odd, such as photocopy onto 11" by 17" paper, they are less likely to be used—better to reformat to 8 ½" by 11" or 14" paper.

Color and Contrast:
- The best contrast for printed material is black or dark ink on yellow, white, or light pastel paper.
- Do not use green, blue, purple, red, or gray paper.
- It is harder to read dark ink on a dark background or light ink on a light background. Light on dark may work.
- It is harder to read text on paper that has lines, pictures, or designs, patterns, multiple colors, or shading.

Other considerations:
- A few or few dozen 18 point or larger bulletins, newsletters, outlines, or song sheets on 8½" by 11" (or 8½" by 14") folded paper are not costly.
- Slightly larger print may help everyone, but 10% or more of the population need true large print.
- When distributing tracts, the Gospel of John, or other literature, some material, at least some of the essentials and contact information should be in large print.

- If creating 100 bulletins each week is typical, add 10 in 18 point Arial font, or make 90 in 12 point and 10 in 18 point. Add more as needed.
- It may take time before people become aware of and comfortable using large print materials. They need not be wasted. Those who read smaller print can use the large print copies.
- Provide paper copies of projected material to people who might not standup, but whose view may be blocked by those standing in front of them.
- Unless we talk about it people may assume their squinting and pain are just their own problem, even though it may be caused by our misuse of technology.
- With a TV, video, or DVD presentation, include a reasonable amount of readable, captioned text and audio narration. It can help everybody.
- A church library can have a large print section with Bibles, hymnals, devotionals, and other large print Christian books.
- Publishers such as, Gale Cencage, Walker, and others, adhere to the standards of the National Association for Visually Handicapped (NAVH) which became part of the Lighthouse Guild in 2010. Christian audio books are also a good addition to a church library.

Several digital software systems have Bibles and reference works with text that can be enlarged. These include the free Bible software program, e-Sword (www.eSword.net). Many types of digital books are available to download for free from the Gutenberg project (www.gutenberg.org/catalog). The Bible software program Logos, www.logos.com, (Faithlife) is for sale. The book texts can be enlarged, but the controls to use them may be small. Kindle (mobi) and Nook (epub) books have some degree of accessibility.

2) Signs: Signs on roads are made to be readable. Signs in grocery stores are large and clear. Signs in churches point to a more important destination and to a greater nourishment.

- All event and information notices on a wall or bulletin board should be in large print with nothing blocking a person's approach to the notice for closer reading.
- A name tag is a sign. Use the largest, clearest text that fits, with the best color contrast, on the typical paper inserted into a 2.5 by 3 inch (6.35 by 7.62 cm) plastic name tag, or whatever size tag everyone wears.

- For guidelines for Braille and other tactile signs see www.access-board.gov.
- Secular laws and common sense require that signs be easily readable for emergency exits, washrooms-lavatories, and fire extinguishers, stairwells, elevators, or a defibrillator unit. Those are not new requirements.
- Text in signs such as door numbers over a door or at a similar distance from the reader should be 3 inches tall by 1.8 inches wide (7.62 by 4.57 cm), with light-colored characters on a dark background.
- Let people know what materials are available and the location, and make extra effort to announce if, or when things are moved, or items are changed, added, or missing.
- Some churches list their accessibility features on a Website, or a visitor packet, and/or have a sign or some other large print handout to let the public know the various ways the church is accessible.
- Readable signs help locate: large print or Braille hymnals, Bibles or audio devices on a shelf, a ramp behind closed doors, or a wheel chair in a closet, or large print song lyrics, bulletins, or sermon outlines.

- Do not only rely on greeters or volunteers to tell people the ways a church is accessible.

3) Bibles: We have a wealth of Bible versions. People who seek to know God need a Bible they can read.

- Bible publishers use their own definitions for the terms, "large," "extra large," "giant," or "super giant print," often without indicating the font size or sizes used. Their terms and font sizes may not be standard large print.
- Bookstore employees who are unaware of these print size variations in the industry may inadvertently mislead customers and some mislabeled items must be taken, or mailed back, or cannot be returned at all.
- The Lutheran Braille Workers produce large print and Braille Bibles. The Jewish Braille Institute distributes the Torah in Braille and in large print. Both organizations, however, may be overwhelmed with requests.
- A complete, one volume, 18 point font, true large print Bible, King James version (KJV), is available from the American or Canadian Bible Society (ABS or CBS), or Holman, or Dake. Holman sells an 18 point font New

King James version (NKJV). All of these would be easier to read if the letter, word, and line spacing were increased.

- Zondervan offers a one volume "super giant print" New International version (NIV) in 16.5 point font.
- At one time, Cambridge University Press published the KJV, the NIV, and the New Revised Standard version (NRSV), each in four volume sets, in 18 point font, which all have good letter, word, and line spacing.
- At one time, Thomas Nelson sold a very readable, 24 point font, three volume KJV.
- ABS sells the easy reading level, Today's English version, New Testament, Psalms and Proverbs, in large print.
- Additional complete versions, in true large print, in one or more volumes, would not be costly to produce. Publishers may respond to requests for new versions, or to re-release the ones they have, but no longer distribute.
- Bible software and eBook devices allow for changing font size for the text, but they need their controls enlarged and better contrast, color, and lighting control. EBooks have some accessibility even on computers. Online HTML

books are more accessible. Alternatives to reading exist, such as, audio Bibles, tape, CD, or Internet streaming, but to hold and to read a print Bible is often preferred over listening to it, or using a computer, smart phone, or eBook.

People may not know the definitions of large print or that the idea to include people with impairments in our feasts of worship and the Word comes from God, from the Bible, and not from secular governments. When churches are accessible we maintain relevance and credibility in our community and we may lead the nations.

4) Computer projection: Microsoft (2014) suggests a one-inch projected letter (72 point font on the computer screen) is readable at a distance of 10 feet. A two-inch letter (144 point font on the computer) is readable at 20 feet. A three-inch letter (216 point font on the computer screen) is readable at 30 feet. Expanding line, word, and character spacing helps too.

The readability of projected text differs due to the dimensions of a room and the distance between the projector and screen, and/or a projector's power. Each auditorium may have different seating, screen size, or ambient lighting issues. Just using a stronger, more powerful projector is not always the best solution.

- Creative layout is fun, but when the purpose is to communicate and have people engage the projected or LCD message, make it as readable as possible. Aim for 95% of the population, not just 51% or 68%.
- In addition to using clearer, universal design for projected slides or for a Liquid Crystal Display (LCD), also make some large print paper copies available.
- Once this is learned, it becomes easy to do.

Slides:
- MediaShout or ProPresenter and most other programs are like Microsoft PowerPoint or Corel Presentations in that these programs all use text boxes on a digital slide.
- Use a slide with only one text box and expand the box to fill the slide.
- Use about 15 words per slide, with a slight margin all around, for song lyrics, or sermon notes, scriptures, or an outline, or a message.

- Instead of 50 words on a slide or a 10 point outline on one slide, enlarge the font and use three or more slides, each with about 15 words.
- Keep It Simple Saint.
- Some slides are pictures of text (.jpeg), not real text. The picture can be enlarged to fill the entire slide or the slide can be reformatted as a text slide in order to enlarge the font size. Take a few minutes to rekey it and save the song or outline for improved readability.
- Text from files or an Internet service that automatically go into a list of slides may be too small and need to be rekeyed into a larger font. This is not difficult to learn and do.

Text:
- Instead of using a font size number such as 72, 144, or 216 point font (1, 2, or 3 inches) as standard, fill the entire screen not just part of it, with about 10 to 15 words per slide, with a slight margin all around.
- Using 15 words, is about enough words for a meaningful English phrase or sentence.
- Slightly more or fewer words (12 to 18) per screen can be used so that the portion on each slide is meaningful and makes sense.

- Other languages may have other standards for projecting a meaningful phrase or sentence.
- This also applies to TV, video, or DVDs that include text and helps both seeing and reading.
- Use the font tool to enlarge the font size. Choose a clear, sans serif font type, such as Arial. Do not use fancy fonts such as cursive script. Do not use appearance affects such as all caps, italics, outline, bold, or shadow.
- Some people like bold text, but usually the spacing must be expanded, otherwise it can seem that the letters, words, and lines blend and blur into each other.
- Text is harder to read when a slide has pictures, designs, multiple colors, a music staff, video, any movement, or many different font types, or font sizes per slide, or in the same sequence of slides.
- If the text already exists, highlight the text and change the font size and type. It may change automatically, or copy and paste the text with a larger, clearer font onto a new slide or slides.

Color contrast:
- Good color contrast is essential. Yellow or white letters on a plain, medium blue background is

preferred by many people, but Kitchel (2008) recommends dark letters on a light pastel background. (See this book's cover for an approximate, but inexact example.)

- View a color contrast wheel Online to see the strongest color contrast choices.
- Color laser pointers are invisible to people who are color blind (most often men).
- Read out loud (ROL), do not just point.
- Do not use white on black or the reverse which creates a glare that feels like staring into a headlight.

Other considerations:

- The purpose is not to simplify, dumb down, or limit the message. The purpose is to project on each slide a balanced chunk of words that conveys meaning, maybe a whole sentence, or a phrase, but no more than three or four lines, with a font clear and large enough to fill the screen with about 10 to 15 words, and a slight margin all around.

- Do not wear yourself out changing all slides at once. A few at a time is a good way to begin.

• Consistently make known in an accessible way that paper copies of projected material are available and where they are regularly located. People are unlikely to ask for these or other accommodations.

5) Internet for Computers and Mobile devices: Internet sites should be accessible to people who do not see or hear well, or who have cognitive or ambulatory limitations. Websites may be used without sound, or color, or in high contrast mode, or with the screen magnified, or with no visible screen just a screen reader, or without a mouse or other focus indicator with voice control only, or a combination of these. The site or apps may be used with accessibility features for Microsoft or Apple, or for Android, IOS, or Windows smart phones, such as Talkback, or Voice Over, or Zoomtext, or JAWS, or using a Braille Display or Notetaker.

• Use W3C Web Content Accessibility Guidelines (WCAG) 2.0, AA, (see (www.w3.org/WAI/intro/components.php)
• Compare your site to accessible Web sites, such as, the American Foundation for the Blind

(afb.org) or lighthouseguild.org, or the National Federation of the Blind (nfb.org), or the American Council of the Blind (acb.org), or joniandfriends.org, or www.churchinwales.org.uk.

- Visit the Google accessibility site,
  www.google.com/accessibility/ and
  the Apple accessibility site,
  http://developer.apple.com/accessibility/.
  See dequeuniversity.com in the USA or usabilitygeek.com in Europe.
- Provide alternative text (alt tags) for images, video, and sound, but not a "Text Only" site which often is not kept up to date.
- A link needs a label and do not use pop-ups.
- Do not use tables for the purpose of layout.
- Design first for mobile devices, tablets, and smart phones, and for apps and what is posted with them. The book Mobile First by Luke Wroblewski (2011) explains why it is easier to begin by building small and then going larger.
- Test the site initially and as it grows with automated Internet evaluation software, such as FireEyes with aXe extension from deque.com.
- These software programs are not 100% perfect so people with disabilities who use various assistive technologies need to periodically test the site.

- Other automated testing services available online are, WAVE – Web Accessibility Versatile Evaluator, by WebAim, Web Accessibility Checker, and AChecker - Accessibility Checker, and Amaze by deque.com.
- Make sure the elements of a Web site will remain visible, usable, and not overlap when the screen is enlarged by pressing the "ctrl" and "+" keys together (zoomed).
- In addition to allowing resizing of text, allow the text to "wrap" or "reflow" without scaling down so that zooming enlarges the text as in a word processor or reflowable eBook, without extending it beyond the screen sides. This will minimize tedious, dizzying, and tiring sideways scrolling.
- Set up PDF text so it can be copied, enlarged, and read in a word processor. PDF documents do not reflow when "zoomed in," or enlarged. They overlap the screen sides and require tedious, dizzying, tiresome, sideways scrolling.
- If pictures are used for visual aesthetics or "eye candy," and contain no content or directional information, include a "skip to content" option.
- Use good semantics. The meaning of headings, titles, and labels is crucial to someone who is not using the picture to understand the message.

- Use more than color or contrast as information, or direction indicators, or for navigation.
- Three screens with about 15 words each, with two or three lines per screen, are easier to read than one screen with 45 words, or five or more lines. Make the slide small print, then enlarge, divide, and copy and paste it to three slides.
- Include captions and transcriptions with video and audio multimedia.
- Set the browser's e-mail default font to a 12 point, sans serif font, such as Arial, without compression.
- A church's Website, Facebook page, or other social media should be accessible. Text is hard to read with small print on top of pictures or textured wallpaper.

We learned to stop smoking, at least in church, and we learned to obey the rules of the road when driving. Instead of driving people away from church, we can learn to build accessible Websites which reflect Christ. We can include those who do not see, hear, or read well, so they see, and hear, and understand the Word of God. People addicted to cell phones or computers can be set free!

6) Hearing Assistance: A hearing Loop, an FM system, and an Infrared system reduce background noise in a hearing aid, making it easier for people to hear clearly, and allow lowering the overall volume to a comfortable level for everyone.

- The cost for any of those 3 hearing systems is between $1,000 and $2,000. Additional personal receivers to connect to a system for those who do not have hearing aids range from $50 to $100 each or more.
- Installing the wire for a Hearing Loop is not hard, but it is easier in new construction. It may cost $1,000 or more if a contractor is hired to retrofit an auditorium.
- One amplifier for a hearing loop covers a room or an auditorium with a parameter up to 350 feet (107 meters) or 87 feet (27 meters) on each of four walls.
- The cost, of about $2,000 for a hearing assistance system, can be added to the budget for a remodelling project, a sound board, piano, synthesizer, or the cost of a missions trip.
- People with sensory impairments or limited reading ability are the largest mission field at home or abroad.

- About half the people in America who need a hearing aid cannot afford one and not all hearing aids have a telecoil that can receive from a hearing loop, so individual assistive listening devices are also needed.
- An FM system for use with up to four headsets may cost $800 for one unit that connects to a sound system.
- Instead of, or in addition to a hearing system, churches may lend individual assistive listening devices (ALDs) for each church service that cost $25 to $100 each. They help those without a hearing aid, but without a hearing system they amplify all background sounds.
- Post signs to indicate what system is available and check that it is turned on and works.
- Real-time captioning of audio material is imperfect, but with a good transcriptionist it may help.
- A Telecommunication Device (phone) for the Deaf (TDD) for a church office may cost $300.
- The American Sign Language Interpreter Network (www.aslnetwork.com/services/video-relay-interpreting) offers video relay sign language interpreting using high speed internet, a computer, and a USB video camera.

- Lower the key (pitch). Transpose a song from the key of D, to D flat, or C, or B, etc., to make it easier to hear. Songs can be transposed with a synthesizer transpose button or guitar capo. Internet and computer programs can transpose sheet music to be printed.
- Singers and speakers need to clearly articulate the consonants of words.

The headings and 80+ points in these seven bulleted lists can be the basis for an ICT accessibility audit. Also audit the accessibility of activities, architecture, and people's knowledge about and willingness to be an accessible church. This is teaching the Bible. Use audit checklists that refer to specific behaviours and terms related to the environment and activities of a church.

All this ICT accessibility may be too much to apply all at once. It does not cover architecture accessibility for people with mobility impairments (found at www.access-board.gov), which in older buildings can be a challenge. It also does not cover access for other special activities (which might be found on www.ada.gov or by contacting specific advocacy groups.)

7) Tactile Writing: The least expensive Braille Embossers cost between, $2,000 and $4,000 and include the software that converts MS Word into Braille. Some software and embossers can print Moon tactile letters as well as Braille. See: www. afb.org/prodBrowseCatResults.asp?CatID=45 for a (2014) discussion and list of Braille Embossers.

- Interpoint Braille uses both sides of the paper.
- Interlinear Braille prints Braille dots and inked text so a person with sight can read the text.
- The special paper may cost $24 for 500 sheets.
- Braille books require more shelf space.
- Moon writing is not widely used. For more information see: www.moonliteracy.org.uk/.
- Instead of buying and learning to use a Braille embosser, some churches e-mail their printed material to a Braille print service provider and receive it via the post office.
- Most large cities have multiple sources to contact for Braille transcription.
- Key concerns when outsourcing the work of Brailling are, timely, reliable, accurate service.

- Contact www.optasiaministry.org to request a library of Bibles and Bible reference works on a DVD for use with computer screen readers or Braille notetakers.

Less technology may be better for the quality of sound or visuals and for the group experience. Leave the stage. Smaller groups do not have to act like TV or mega churches or college classes. Consider teaching a lesson or song antiphonally, that is, responsively speaking or singing, with a congregation repeating a line after the leader. This method is often used to impart our oral tradition or when there is a power outage!

Whatever the style or size of a church and whoever may be the intended audience, we can easily learn to use the ICT we own, that is, our computer printing, projection, LCD, photocopy, smart phones or smart boards, Websites, e-mail, social media, and sound systems -all- in inclusive ways. Disability discrimination is illegal where we live. We represent the one true God who cares about those who do not see, hear, or read well.

Unnecessarily excluding people from church is harming them. Misuses of technology may also cause physical harm. A decibel (db) meter, or Sound Pressure Level (SPL) meter, able to measure 30 db to 130 db, can cost $10 to $60. There are SPL apps for cell phones. The accuracy of different sound meters may vary by as much as 20 db. Exposure to 100 db of sound should be avoided to avoid damage to hearing.

Approximate font sizes   John 3:16 (NKJV)

(Arial 12 point font)   For God so loved the world,

(14 point)   that He gave

(16 point)   His only begotten Son,

(18 point)   that whoever believes in Him

(20 point)   should not perish

(24 point)   but have

(36 point)   everlasting

(72 point)   life

11. Conclusion

Jesus asked the leaders of a synagogue if it were lawful to do good or to do harm on the Sabbath. They were silent. Their highest standard was not whether an activity helped anyone with or without a disability. What mattered to them was what activities the law and their traditions permitted on a Sabbath day for those who worshiped God. Jesus was grieved and angered at their hardness of heart (Mark 3:1-6). He pointed to the standard of loving God and their neighbor, but they rejected it and him.

The presence and perspective of people with severe impairments exist in scripture. The topic and people warrant inclusion in any size or style of church. They are part of the Bible and life. Our hope of resurrection is based on the cross, and there is no crucifixion, no Gospel without impairment and disability. We cannot fulfill the Great Commandments or the Great Commission without people with impairments. We do not have religious freedom as a church or as a nation if we block and deny access to worship to millions of our neighbors and our citizens.

Religious institutions benefit from God's gifts of government and law. The churches' environment and context include Biblically based laws restricting how information and communication technology is used. Such regulation is normal and beneficial. Disability discrimination is a crime in many countries. Nevertheless, the laws of any country are only as good as their enforcement and the moral character of its citizens. We, the Church are called to be the light of the world. Intentionally being an accessible church means we pro-actively remove barriers, implement universal design, and make reasonable accommodations and adjustments available before anyone requests them and that we talk about this because they are central to who Jesus is and to the Gospel we believe.

We cannot assume who is, or will become a Christian, or who wants or needs to participate in church gatherings. Wherever a church needlessly disables, hinders, and harms people by blocking their participation she can confess, repent, break the silence, and correct the misuse of technology. After becoming accessible there will still be a need for good works by individuals and by groups. A charity and benevolence ministry will continue to be necessary parts of church life. Various forms of competition can still exist, but we will better represent God and the good news if we create accessible church ministry activities, including spaces, literature, and evangelistic, and missionary outreach, by the inclusive use of information and communication technology.

The Gospel, the good news extended to all, includes the message that a relationship with God is possible for a greater range of people. That includes those whom our Lord Jesus Christ highlighted by words and deeds and specifically told us to bring to our feasts of worship and the Word. Jesus made a way for people who could not see, hear, talk, walk, or think well and so can we. Prepare the way for the people (Isaiah 57:14).

The abundant life Jesus offers us (Jn. 10:10) and the possibilities of a high quality of life with a severe impairment are taught, in part, by the church's example. The messages we try to communicate are important now and for eternity. We can counter a culture of death by creating a community of life. Let us teach a greater understanding of human variation in our classrooms, pulpits, and Christian media and become inclusive by demolishing man-made barriers. We may have failed trying to lead in other areas because of neglecting this one very obvious and central area. With the Word and the Holy Spirit as our guide, we can learn to do this and lead the nations. We are a church.

More than 300 years ago Isaac Watts wrote, "See how we trifle here below, fond of these earthly toys: our souls, how heavily they go to reach eternal joys." So much eternal joy eludes us as we ignore our misuse of information and communication technology. Impairments, like death, are a part of life. We fool ourselves if we think silence about it and about disability discrimination are just ways we try to reach young people. We are offering them a Christianity without Christ; without him who helped people see and hear and understand the Word of God; –without the Christ whose feet and hands were disabled for us on the cross. We are called to be witnesses of his resurrection. Our silence and misuse of technology must be challenged for the good of those being hurt and for those who cause the pain and for all who need our message.

> When I survey
> the wondrous cross
> on which the Prince
> of Glory died,

> My richest gain
> I count but loss
> and pour contempt
> on all my pride.
>
> When I survey the Wondrous
> Cross. By Isaac Watts, 1707,
> Public Domain.

The 27 words of the hymn text above can be put in two slide boxes with 13 and 14 words. Presenting and preserving the meaning and the impact of a text requires more than only counting up to 15 words. Accessibility also requires and allows thoughtful consideration of the content.

* * * * * * * *

Tell of the cross
where they nailed Him
writhing in anguish
and pain;

Tell of the grave
where they laid Him
tell how He's living again.

Love, in that story
so tender,
clearer than ever I see;

Stay, let me weep while
you whisper, "His love paid
the ransom for me."

Tell Me the Story of Jesus.
Fanny J. Crosby, 1915, Public domain

The 50 words of the hymn verse above are clearer when projected on four slides of 13, 13, 11, and 13 words.

Amazing grace
how sweet the sound
that saved
a wretch like me!

I once was lost
but now am found,
was blind but now I see.

Amazing Grace.
John Newton, 1779,
Public Domain

These 26 words, even though familiar, could be enlarged onto two slides with 12 and 14 words.

* * * * * * * *

Whether or not anyone who does not see, hear, or read well, asks for access, or attends, or will attend our church, we can set up ICT all the time so more people will find it easier to see, read, hear, and understand songs, hymns, messages, outlines, scriptures, bulletins and announcements.

We can print and project with our computers and use our sound systems, cell phones or smart boards, and the Internet and social media, all accessibly and be inclusive.

We can **turn barriers into bridges.**

12. References

Action On Hearing Loss. (2014). Hearing Loss. Retrieved January 11, 2015 from:
www.actiononhearingloss.org.uk/.

Albrecht, G., & Devlieger, P. (1999). The Disability Paradox: High quality of life against all odds. Social Science and Medicine, 48(8), 977-988.

American Foundation for the Blind (AFB). (2012). Summary, National Center for Health Statistics for U.S. adults, National Health Interview Survey (NHIS), 2012. Retrieved July 24, 2014 from: www.afb.org.

Americans With Disabilities Act of 1990, As Amended. (2009). (ADAAA, 2008) Retrieved August 11, 2014 from: www.ada.gov/pubs/adastatute08.htm.

Accessibility for Ontarians with Disabilities Act Alliance [AODA Alliance]. (2015). Analysis Of Premier Kathleen Wynne's December 23, 2014 Letter and Economic Development Minister Brad Duguid's December 8, 2014 Letter to the AODA Alliance. Retrieved, January 29, 2015 from: www.aodaalliance.org/strong-effective-aoda.

Anderson, M. [the ICT evangelist] (2013). Perfect ICT every lesson. Independent Thinking Press, an imprint of Crown Publishing, Wales, UK.
www.Independentthinkingpress.com.

Barnes, C. (1996). Disability and the myth of the independent researcher. Disability and Society, 11(1), 107-110.

Bagenstos, S. (2004). Has the Americas With Disabilities Act reduced employment for people with disabilities? Berkeley Journal of Employment & Labor Law, 25, 527-563.

Bickenbach, J. (2000). The ADA v. the Canadian Charter of Rights: Disability rights and the Social Model of disability. In L., Francis, & A. Silvers (Eds), (2000) Americans with disabilities: Exploring implications of the law for individuals and institutions. New York: Routledge.

Blanck, P. (2014). eQuality: The struggle for web accessibility by persons with cognitive disabilities. Behavioral Sciences & the Law, 32(1), 4–32.

Brault, M. (2012). Americans With Disabilities: 2010, Current Population Reports, P70-131, U.S. Census Bureau, Washington, DC. (2012 Estimates based on the Survey of Income and Program Participation [SIPP]).

Calvin, J. (1846, 1997). Institutes of the Christian religion. Translation of: Institutio Christianae religionis.; Reprint, with new introduction. Originally published: Edinburgh: Calvin Translation Society, 1845-1846. (III, x, 3). Bellingham, WA: Logos Research Systems, Inc.

Canadian Association of the Deaf. (2012). Statistics On Deaf Canadians. Retrieved January 12, 2015 from: www.cad.ca/statistics_on_deaf_canadians.php.

Canadian Hearing Society (CHS). (2013). Prevalence of Hearing Loss. Retrieved January 15, 2015 from: www.chs.ca/facts-and-figures.

Canadian Literacy Network [CLN]. (2015). Literacy Statistics (from IALSS 2003). Retrieved January 12, 2015 from: www.literacy.ca/literacy/literacy-sub/.

Centers for Disease Control and Prevention (CDC). (2021). Disability Impacts All of Us Infographic. Retrieved August 20 2021 www.cdc.gov/ncbddd/disabilityandhealth/infographic-disability-impacts-all.html

Charlton, J. (2000). Nothing About Us Without Us: Disability Oppression and Empowerment. University of California Press; New Ed edition.

Church attendance estimates. (2004-13). Weekly statistics. Retrieved: https://en.wikipedia.org/wiki/Church_attendance

Davie. A. (1997, 2016). That All May Worship, An Interfaith Welcome to People with Disabilities. Ginny Thornburgh (Ed.). National Organization On Disability

(NOD), and The American Association of People with Disabilities (AAPD), Interfaith Disability Advocacy Coalition: Washington DC., 7th edition (2016), Retrieved May, 2016, from, http://www.aapd.com/wp-content/uploads/2016/03/That-All-May-Worship.pdf

Dickens, Charles (1853). Bleak House, Bradbury and Evans, London.

Dickens, Linda (2008). Regulating for equality in employment – is Britain on the right road with the Single Equality Bill? Paper presented at the Third International Conference on Interdisciplinary Social Sciences, Prato, Italy. Retrieved January. 29, 2015 from:
www2.warwick.ac.uk/fac/soc/wbs/research/irru/publications/recentconf/ld_-_prato_regulating_for_equality_in_employment.pdf.

Dickens Linda (2014). The Coalition government's reforms to employment tribunals and statutory employment rights —echoes of the past. Retrieved Jan. 30, 2015 from: http://onlinelibrary.wiley.com/doi/10.1111/irj.2014.45.issue-3/issuetoc.

Edersheim, A. (1896, 2003). The Life and Times of Jesus the Messiah. Longmans, Green and Co., New York, London, and Bombay. Vol. 2, Chapter 15, pgs. 234-237. Bellingham, WA, Logos Research Systems, Inc.

Employment and Social Development Canada. (2015). Indicators of Well-being in Canada. Learning - Adult Literacy. First Results from the Programme for the International Assessment of Adult Competencies (PIAAC), Table B.4.1 Literacy and numeracy - Averages and proficiency levels of population aged 16 to 65 in ALL and PIAAC, Canada, 2003 and 2012, Catalogue no. 89-555-X, Ottawa, 2013.

Employment and Social Development Canada. (2015). Indicators of Well-being in Canada. Canadians in Context - People with Disabilities. Canadian Survey on Disability (CSD, 2012). Retrieved January 10, 2015 from: www.hrsdc.gc.cawww4.rhdcc.gc.ca/indicator.jsp?&indicatorid=40.

Feinberg, J., & Feinberg, P. (1993). Ethics for a Brave New World. Crossway Books, Wheaton IL. Bellingham, WA, Logos Research Systems, Inc.

Frank, J. (1999). Internet or nyet? Some internet prospects and problems in rehabilitation education. *Rehabilitation Education,* 13(2), 163-172.

Frank, J. (2000). Requests for large print accommodations by persons who are visually impaired. Journal of Visual Impairment and Blindness, 94, 716-719.

Frank, J. (2002). What are we teaching about the ADA and what is missing? Re:view, 33, 171-181.

Frank, J. (2006). A survey of the Americans With Disabilities Act (ADA) accommodation request experience of persons who are blind or who have a severe visual impairment. Mississippi State University, MS: Rehabilitation Research and Training Center on Blindness and Low Vision (RRTC-BLV). (now called the National RTC-BLV.)

Frank, J., & Bellini, J. (2005). Barriers to the accommodation request process of the Americans With Disabilities Act. Journal of Rehabilitation, 71(2), 28-39. Or: www.questia.com/read/1G1-133317582/barriers-to-the-accommodation-request-process-of-the.

Frank, J., & Stephenson, M. (2013). Let's end disability discrimination in church. The Banner, 148(10), 36-37. Or: www.thebanner.org/features/2013/09/let-s-end-disability-discrimination-in-church.

Gallup Poll (2010). Church Attendance. Retrieved Sept. 4, 2014 from: www.gallup.com/poll/166613/four-report-attending-church-last-week.aspx.

The Hearing Loss Association of America (HLAA). (2014). Who We Are. Communication Access Technology Survey. Retrieved Sept 2014 from: www.hearingloss.org/ sites/ default/files/docs/HLAA_CommAccessTech_Survey _Results.pdf

IDEA: The Individuals with Disabilities Education Act. (1975). Retrieved August 20, 2021 from https://sites.ed.gov/about-idea/.

Indicators of Well-being in Canada. (2012). Learning - Educational Attainment. Retrieved, March 28, 2016 from: http://well-being.esdc.gc.ca/misme-iowb/.3ndic.1t.4r@-eng.jsp?iid=29.

Johnson, M. (2003). Make Them Go Away: Clint Eastwood, Christopher Reeve & The Case Against Disability Rights, Advocado Press Inc.

Johnson, M., & Frank, J. (2004). Stories of accommodation gone wrong. The Ragged Edge, Advocado Press. Retrieved August 5, 2014, from: www.raggededgemagazine.com/ADAaccores1004.html.

Kagle, J., & Cowger, C. (1984). Blaming the client: Implicit agenda in practice and research? Social Work, 29,347-351.

Kaye, S. (1998). Is the Status of People with Disabilities Improving? Disabilities Statistics Center, University of California, San Francisco Abstract 21. Retrieved Aug. 13, 2014, from,http://dsc.ucsf.edu/publication.php?pub_id=7.

Kistemaker, S. (1986, 2001). Vol. 14: New Testament commentary: Exposition of James and the Epistles of John. New Testament Commentary (260). Grand Rapids: Baker Book House. Bellingham, WA, Logos Research Systems, Inc.

Kitchel, J. (2004). APH guidelines for print document design. American Printing House for the Blind. Retrieved Aug. 11, 2014 from: www.aph.org/edresearch/lpguide.htm.

Kitchel, J. (2008). Color and text guidelines for the development of Power Point, computer and Web page presentations and text applications for audiences that may include persons with low vision. The American Printing House for the Blind, Louisville, KY. Retrieved Aug. 11, 2014 from:
www.aph.org/edresearch/ppt_guidelines.ppt.

Koster, S. (2005). Calvin Institute of Christian Worship, 2003, Projectors in worship: Survey summary: Results of research on the use of technology in worship. June 25, 2005. Retrieved, July 25, 2014 from:
http://worship.calvin.edu/resources/resource-library/projectors-in-worship-survey-summary/.

Matson, F. (1990). Walking Alone and Marching Together: A History of the Organized Blind Movement in the United States, 1940-1990. National Federation of the Blind.

Microsoft. (2014). Tips for creating and delivering an effective presentation - PowerPoint. Retrieved August 11, 2014, from: http://office.microsoft.com/en-us/powerpoint-help/tips-for-creating-and-delivering-an-effective-presentation-HA010207864.aspx.

Moss, K., Burris, S. Ullman, M., Johnsen, M., & Swanson, J. (2001). Unfunded mandate: An empirical study of the implementation of the Americans with Disabilities Act by the Equal Employment Opportunity Commission. Kansas Law Review, November, 50(1). 1-110.

National Association of Evangelicals (NAE). (2015). Code of Ethics for Congregations and Their Leadership Teams. retrieved Oct. 7, 2015 from: http://nae.net/code-of-ethics-for-congregations/

National Health Statistics [NHS]. (2013). Vision impairment. Retrieved Jan 12, 2015 from: www.nhs.uk/conditions/Visual-impairment/Pages/Introduction.aspx.

National Highway Traffic Safety Administration. (2014). Fatality Analysis Reporting System Retrieved August 29, 2014 from: www-fars.nhtsa.dot.gov/Main/index.aspx.

National Institute on Deafness and Other Communication Disorders. (2014). Quick Statistics. Retrieved December 12, 2014, from; www.nidcd.nih.gov/health/statistics/pages/quick.aspx.

National Institute of Literacy. (2013). Adult Literacy Survey. United States Department of Education. Retrieved July 24, 2014 from: www.statisticbrain.com/number-of-american-adults-who-cant-read/.

National Literacy Trust. (2011). Literacy in Britain. Retrieved Jan. 12, 2015 from: www.literacytrust.org.uk/adult_literacy/illiterate_adults_in_england.

O'Brien, R. (2001). Crippled Justice: The history of modern disability policy. Chicago: The University of Chicago Press.

O'Connor, C. (2013). Where is the accessible, tactile currency? Retrieved July 22 2014, from; www.pdrib.com/blog/where-is-the-accessible-tactile-currency/.

Office of National Statistics [ONS]. (2012). Statistical bulletin: Adult Health in Great Britain 2012. Retrieved Jan 12, 2015 from: www.ons.gov.uk/ons/rel/ghs/ opinions-and-lifestyle-survey/adult-health-in-great-britain–2012/stb-health-2012.html.

Potok, A. (2002). A matter of dignity. New York: Bantam Books/Random House.

Rogers, E. (2003). Diffusion of Innovations (5th ed.) Simon and Schuster.

Rubin, S., & Roessler, R. (2001). Foundations of the Vocational Rehabilitation Process. (5th ed.) Austin, TX: pro-ed. Chapter 2 & chapter 4.

Ryan, W. (1971). Blaming the Victim. New York: Pantheon Books.

Sanchez, W., & Fried, J. (1997). Giving voice to students' narratives. College Teaching, 45 (1), 26.

Shapiro, J. (1994). No Pity: People with Disabilities Forging a New Civil Rights Movement. Broadway books.

Singleton, P. (2009). Insult to Injury: Disability, earnings, and divorce. Syracuse University, Depart, of Economics. Retrieved July 22, 2014, from: http://papers.ssrn.com /sol3/papers.cfm?abstract_id=1553246.

Stade, J. (2014). Eye Health Data Summary: A review of published data in England. Royal National Institute of Blind People [RNIB]. Retrieved January 12, 2015 from: www.rnib.org.uk/sites/default/files/Eye_health_data _summary_report_ 2014.pdf.

Statistics Canada. (2014). Canada's population estimates: Age 2014. Retrieved Jan. 12, 2015, from: www.statcan. gc.ca/daily-quotidien/140926/dq140926b-eng.htm.

Statistics Canada. (2014). Population by year by province and by territory. Retrieved January 12, 2015 from: www.statcan.gc.ca/tables-tableaux /sum-som/l01/cst01/demo02a-eng.htm.

Statistics Canada. (2010). Participation and Activity Limitation Survey (PALS) 2006, Retrieved Jan 10, 2015 from: www.statcan.gc.ca/pub/89-628-x/89-628-x2010015-eng.htm.

Stumbo, E. (2014). Confessions of a pastor's wife: The church is forgetting us. Retrieved July 25, 2014 from: www.ellenstumbo.com/confessions-pastors-wife-church-forgetting-us/.

Temple, G., with Ball, L. (2012). Enabling Church. www.torchtrust.org, or the Society for promoting Christian Knowledge, www.spckpublishing.co.uk.

United States Access Board. (2002). ADA Accessibility Guidelines (ADAAG), section 4.3 Signage. Retrieved August 14, 2014 from: www.access-board.gov/guidelines-and-standards/buildings-and-sites/about-the-ada-standards/background/adaag#4.30s.

United States Department of Justice. (2012). 2010 ADA Standards for Accessible Design, Retrieved August 11, 2014 from: www.ada.gov/2010ADAstandards_index.htm.

Vander Plaats, D. (2014). 5 Stages: The journey of disability attitudes. Elim Christian Services. Retrieved August 13, 2014 from: www.elimcs.org/news/elim-worship-team-chicago-sun-times.

The World Health Organization (WHO). (2001). International Classification of Functioning, Disability and Health (ICF). Retrieved August 5, 2021 www.who.int/standards/classifications/international-classification-of-functioning-disability-and-health.

Wright, B. (1981). Value laden beliefs and principles for rehabilitation. In S. Regnier & M. Petkovsek (comp.) (1985), 25 years of concepts, principles, precepts. A collection of articles published in rehabilitation literature 1959-1984. (pp 113-116). Chicago, IL: National Easter Seals Society.

Yong, A. (2010). Disability, the Bible and the Church. Eerdmans, Grand Rapids MI, pages. 141-142.

* * * * * * * *

"I long to accomplish a great and noble task, but it is my chief duty to accomplish small tasks as if they were great and noble."- Helen Keller.

* * * * * * * *

"There are many of us that are willing to do great things for the Lord, but few of us are willing to do little things." – Dwight L. Moody.

* * * * * * * *

"If revival is being withheld from us, it is because some idol remains still enthroned; because we still insist in placing our reliance in human schemes; because we still refuse to face the unchangeable truth that, 'It is not by might, but by My Spirit." – Jonathan Goforth.

* * * * * * * *

"Someone asked, Will the heathen who have never heard the Gospel be saved? It is more a question with me whether we – who have the Gospel and fail to give it to those who have not – can be saved." – Charles Spurgeon.

* * * * * * * *

Only one life, t'will soon be past,
Only what's done for Christ will last.
                    – C.T. Studd.

About the Author:

John Jay Frank, an Ordained Minister of the Gospel and a Certified Rehabilitation Counselor for many years, has had training and experience in Reading Education, Music, Biblical Studies, and Information and Communication Technology. He earned a PhD in Rehabilitation Counseling at Syracuse University in New York (2003).

Beyond personal experience with his own vision impairment, his professional and volunteer work included direct care of people with various severe impairments and research and teaching positions, as well as pastorate, chaplaincy, missionary, and evangelistic activities.

Dr. Frank is the founder of Minstrel Missions LLC. He writes articles and books and records songs and hymns for the encouragement, education, edification, and comfort of the body of Christ.

See: www.minstrelmissions.com. E-mail comments and questions to: minstrelmissions@gmail.com
also see John Jay Frank on Youtube.com
Books are available at amazon.com

4 Books:

Turning Barriers Into Bridges: The Inclusive Use of Information and Communication Technology for Churches in America, Britain, and Canada

Myths, Lies, & Denial:
Christian and Secular Counseling in America

A Minstrel's Notes:
Stories and Sermons on Worship in Spirit and In Truth, and Music Theory and Technique for the Acoustic Guitar.

Come Worship In Spirit and In Truth.

7 Music Recordings (CDs):

For All God's Children
Comfort Ye My People
The Cross Of Life

Hymns of the Church
Hymns and Carols
Hymns of God's Grace
Hymns of Emmanuel

A Music DVD:

Pick'n and Preach'n, (containing 12 songs filmed in Connecticut and California.)

Love one another as I have loved you
(John 15:12).

Go therefore and make disciples
(Matt. 28:18-20).